PILOT'S WEATHER GUIDE

By
LINDY BOYES

NEW YORK

MODERN AIRCRAFT SERIES

A DIVISION OF SPORTS CAR PRESS

MODERN AIRCRAFT SERIES

Edited by Joe Christy

Beechcraft Guide
Parachuting for Sport
Guide to Antique Planes
Cockpit Navigation Guide
Used Plane Buying Guide
Air Traffic Control
Modern Aerobatics
Classic Biplanes
Classic Military Biplanes
Aviation Radio for Pilots
Pilot's Weather Guide

Computer Guide
Guide to Homebuilts
Fighter Aircraft Pocketbook
Bomber Aircraft Pocketbook
Racing Planes Guide
Lightplane Engine Guide
The Piper Cub Story
Cessna Guide
Agricultural Aviation
Your Pilot's License
Instrument Flying Guide

Library of Congress Card Number 62-18923

© 1962 by Sports Car Press, Ltd.

Published in New York by Sports Car Press, Ltd., and simultaneously in Toronto, Canada, by Ambassador Books, Ltd.

All rights reserved under International and Pan American Copyright Convention.

Manufactured in the United States of America by Indiantown Printing, Inc.

Second Printing, July 1965

DEDICATION

To my parents, whose constant interest in their daughter's career in both aviation and writing has made this book possible.

Cover photo was taken by Keith Dennison over Guam.
All photos, where not otherwise noted, taken by the Author.
All charts are by Frank Kettlewell.

ACKNOWLEDGMENT

In fifteen years of flying for fun and business, during which time I acquired a commercial pilot's license and a flight instructor's rating, I picked up the jargon of the weather station and learned quite a bit about what goes on there. I asked a veteran meteorologist to check this book for technical accuracy.

My expert was Edmund P. Norwood, meteorologist in charge of the U. S. Department of Commerce Weather Bureau Station at the Metropolitan International Airport, Oakland, California.

Without going into detail, let me express my deepest gratitude for his invaluable assistance. And also to Ellis D. Pike, of the Weather Bureau Station at Wichita, Kansas.

—L. B.

Contents

Chapter	Page
Introduction	7
1. The Air We Fly In	11
2. What's in a Weather Station?	16
3. The Fate of Fair Weather Pilots	36
4. Fly for Fun and Still Reach your Destination	45
5. Clouds and Where They Come From	57
6. Circulation of Air Masses	75
7. Fronts: Cold, Warm and Occluded	85
8. So You Think You Can Get Out on Instruments!	97
9. *Specialités de la Maison*	113

Which way for best weather? Douglas "Wrong Way" Corrigan has one suggestion, while pilots Lindy Boyes, left, and Margaret Callaway add their own confusing touches. Scene is the start of the All-Woman Transcontinental Air Race, 1951, Santa Ana, California. Some of the weather experiences related in chapters ahead were shared by the two women.

INTRODUCTION

I was a fairly new pilot when I encountered my first experience with "weather." And it was one of those classic situations that are always used as examples in instruction books. Unfortunately, I was way behind in my book reading.

Flying a side-by-side Taylorcraft, I was still riding high in spirits from the fun I'd had landing (illegally) on a county road and taxiing up the driveway to the house that belonged to my aunt and uncle on their ranch in California's San Joaquin Valley. The little plane parked in the driveway stopped every passerby in his tracks, and I was feeling pretty pleased with myself.

Later, with my uncle standing guard a few hundred yards down the road to divert any oncoming traffic, I taxied back out and gaily took off. Oh, I was a hot pilot, all right, in those days. I do not advocate illegal flight activities for anyone.

The day was a light gray everywhere; the sun making its presence known only by dim rays through the high overcast. I flew to the nearby Fresno Air Terminal to check weather en route and at my destination. Home base was the Contra Costa County Airport at Concord, on the northeast side of San Francisco Bay. The weather station personnel told me that as I flew north up the valley the weather would improve. Thus reassured, I headed for home.

The further north I went, the lower I had to fly. The ceiling was just not cooperating with the forecast. At something like 500 feet I was down to following the highway. Visibility was reasonably good. It was the lowering ceiling that was annoying. But determined to get home—I could think of nothing worse than being forced to land just 50 miles from base and having to continue on by bus—I skimmed along at the bottom of the descending ceiling.

I was over the town of Antioch with only a few miles to the airport, just short of a hill that sloped down to the water's edge, when the ceiling and the ground met. It was like flying through a pot of pea soup, and it was definitely not for me. I decided to turn around and beat a hasty retreat.

By the time the plane had completed a 180-degree turn, I was out of the cloud. At this point, my only interest was to get on the ground. I retraced my flight path a couple of miles to what had been a booming U. S. Army base. Now there was only a handful of personnel who were in the process of decommissioning the installation.

One part of the base had been cleared of buildings and only a checkerboard of paved streets remained—a perfect "forced landing" site. After looking the streets over, I selected one and made a normal pattern approach and landing. I had barely come to a stop when an official car roared up.

A captain stepped out of the car as I climbed from the plane. Why was I there, he asked me. I told my story fearing that, at best, I would be arrested for landing on government military property.

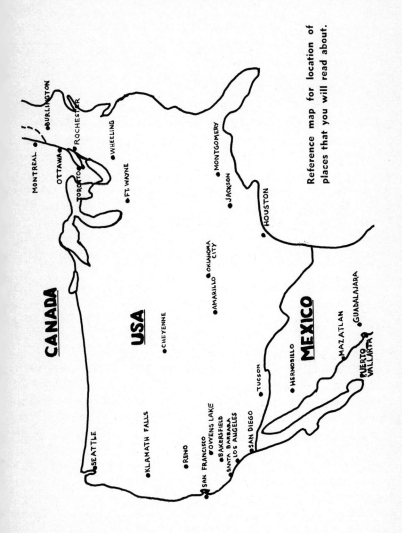

"You're the sixth airplane to use this street for an emergency landing," the captain said, smiling. "I did it once myself."

After we secured the plane, I was driven the few miles to the airport that I couldn't reach by air. Later in the day the weather cleared and the plane was retrieved.

This experience has remained indelible in my flying memory, and I have not pushed my luck that far again. I have since flown in and around almost every kind of weather, but with caution and planning.

Thus we come to the purpose of this book—to introduce the new private pilot to the problems of weather by studying the actual experiences of pilots. Also included are illustrations to help the reader recognize different weather situations, and the information necessary to help you plan and make safe cross-country flights.

Most of all, we want you to enjoy your flying, and we want you to fly safely. Here are a few thoughts to paste in your helmet:

Weather is the biggest bugaboo to flying.

Thunderstorm turbulence can destroy a plane that ventures within its massive form.

Hail can tear fabric from a plane.

The weight of ice can force a plane down to the ground.

Fog can fatally destroy a pilot's perception.

The object lesson is this: you must be constantly aware of weather and what different conditions and situations promise.

You must be aware of your limitations, and fly accordingly.

And, in spite of the foregoing tale, you should always use the facilities and information that the weather station provides.

1. The Air We Fly In

The same air that gives us life also holds us aloft in our man-made flying machines. Just think how important that odorless, tasteless, invisible, gaseous thing really is. It's truly a remarkable mixture.

That air which surrounds the earth is called the *atmosphere*. The atmosphere extends from the earth's surface to an unknown upper limit. Because it is a gas, as it rises it expands and becomes less dense. Half of the entire atmosphere is below 18,000 feet, and three-fourths of the atmosphere is below 36,000 feet. The remaining one-fourth spreads out and up some 500 miles.

Within the atmosphere there is *weather*. Weather is activity of the atmosphere; some of it can be seen, other types felt or measured. Clouds, fog, rain, haze, snow, for instance, can be seen. Turbulence and wind can be felt. Temperature, pressure and humidity can be measured.

In case someone should ask you, the atmosphere is made up of three regions. From the ground up, they are troposphere, stratosphere and ionosphere.

As mentioned, the atmosphere is a gaseous mixture. It contains nitrogen, oxygen, carbon dioxide and water vapor in significant quantities. Water vapor is the one ingredient that changes quantity noticeably in proportion to the other gases. Most water vapor is concentrated below 30,000 feet—where most light plane flying is done.

In addition to the chemical ingredients that compose the atmosphere, there is a great deal of solid matter in the form of particles; these range from submicroscopic size to some big enough to be seen with the naked eye. This matter includes dust, smoke and salt from ocean water. They play an important part in the formation of clouds, fog and precipitation.

With all this matter and water vapor, it is not really surprising that the atmosphere is pretty heavy. It exerts a pressure on the earth's surface at sea level of 14.7 pounds per-square-inch. If the human body didn't contain an equal pressure through body fluids and air from the inside out, the atmosphere's pressure would be a crushing force.

Since, at present, light plane flying is undertaken in the area of the troposphere, we'll end discussion of the atmosphere with these last statistics:

Troposphere extends up some 11 miles at the equator. It tapers towards the poles to as low as 5 miles high. In the troposphere, temperatures generally decrease with altitude, but not necessarily at a constant or uniform rate. There may even be layers where temperatures *increase* with height.

Stratosphere, the next layer, rises from the top of the troposphere to about 50 miles above the earth. Here the temperature changes to constant with increasing altitude.

Ionosphere, the top layer, rises from 50 miles above the earth and is believed to top off at 500 miles. The Aurora Borealis, or Northern Lights, occurs here.

What's In a Temperature?

To achieve any temperature there must be heat, or lack of it. As pertains to light plane flying, temperatures are significant. The sun heats the atmosphere and the earth. The fact that this heating business is not uniform over the earth results in weather. This affects your flying.

Sun (or solar) heat is known as radiation because when it reaches the earth's atmosphere, some of it is reflected (radiated) back into space. Some of the solar radiation reaches the earth's surface before it is reflected back into space. Through evaporation from oceans, lakes and rivers, and from transpiration from vegetation, the lower atmosphere contains moisture. This moisture, in the form of invisible water vapor, produces clouds through condensation.

Clouds limit the amount of incoming solar radiation, as already described. The more cloud coverage there is, the less heat reaches the earth.

The thermometer is the standard instrument for measuring heat. There are two scales in common use—Farenheit and Centigrade. This is as good a place as any to compare them, and to give you the handy system for conversion from one to the other.

Based on the high degree of the boiling point of water and the low degree of the melting point of ice, the scales are:

Farenheit	Centigrade	
212°	100°	—water boils
32°	0°	—ice melts

To convert: $C = 5/9 \ (F - 32)$
$F = 9/5 \times C + 32$

Example; if present temperature Fahrenheit is 62°, you can find the equivalent temperature in Centigrade this way:

$C = 5/9 \ (62° - 32)$ *Ans.* 16.6°C

Using 16°C, reverse the problem to find the Fahrenheit equivalent:

$F = 9/5 \times 16 + 32$.   *Ans.* 60.8°F

Heat Transfer

Terms that have to do with heat transfer may sound familiar . . . conduction, radiation, convection and advection. Their meanings may not seem quite as familiar.

Conduction: heat is transferred from one object to another when the objects are in contact. Or, heat is transferred from one part of an object to another part. To illustrate; if you have ever stirred the contents of a saucepan on the stove and felt the handle of the spoon become warm, warmer, hot and, finally, too hot to hold, you have experienced conduction type of heat transfer. A similar process takes place in the atmosphere when heat is transferred from the earth's surface; the atmosphere in contact with the earth's surface transfers heat by conduction from the earth to the atmosphere.

Radiation: in a wave motion similar to radio or light waves, heat is transferred without benefit or aid of a medium or contact. This is the process by which the sun heats the earth.

Convection and *Advection* are terms applied to mass movements of air through which there is a transfer of heat. When the movement is vertical—hence the heat is transferred vertically—it is called convection. When the process is horizontal, it is advection.

Temperature decreases with altitude at an established rate for given circumstances. This is called the temperature lapse rate.

There are occasions when the normal temperature decrease with altitude is reversed. This situation is a temperature inversion. An inversion at the surface is normally caused by the advection (horizontal movement) of warm air over a colder surface. It also occurs at night when the air in contact with the earth's surface cools rapidly.

Restrictions to visibility can be expected with surface temperature inversions . . . fog, haze, smoke, low stratus clouds.

Upper air inversions are generally caused by the advection of warm air over cold air, of colder air under warmer air, or by air that is warmed by sinking to a lower level. There is usually some type of visibility restriction along with smooth air.

Stability is a word that is used frequently in reference to air. It refers to the ability of the atmosphere to help or hinder the development of vertical air currents. The usual terms are "stable" or "unstable."

Briefly, stable air tends to remain at a given altitude, and if it is forced to move will return to its original altitude. It tends to suppress vertical currents. The air is smooth, but visibility may be restricted.

On the other hand, unstable air continues to move once it is displaced and does not return to its original position. It aids the development of vertical air currents (with assistance from another source such as surface heating and lifting over mountains). Poor flying weather is associated with unstable air . . . thunderstorms, turbulence, hail, gusty surface winds, showery precipitation.

Temperatures over land and sea vary from day to day and season to season because the earth is tilted $23\frac{1}{2}$ degrees from the vertical with reference to the sun, and is spherical in shape. This puts the equator and temperate regions in the position to receive more concentrated portions of direct radiation rays of the sun in all seasons of the year. The lowest surface temperatures are near the poles.

Two other important factors that affect the horizontal temperatures are (1) distribution of land and water masses, and (2) distribution of mountains.

The greatest and most rapid temperature changes are over land masses. Mountains are good obstacles to block the flow of warm and cold air. In the United States, the Rocky Mountains —which run north and south, approximately—prevent the cold air from Canada and Alaska from moving west when it hits the eastern slope of the mountains. This explains the relatively warm temperatures year round in the states west of the Rockies. In Europe, the Alps—which lie east-west—perform the same service for the countries on the Mediterranean.

So much for the atmosphere. Now let's move on to its practical applications for the pilot.

2. What's in a Weather Station?

Do you start off on a cross-Country flight with a visit first to the local weather station? If so, you've seen charts of different kinds—synoptic weather, winds aloft, long range forecasts—and teletype machines that click away. You may think it is foolish to try and decipher this complex maze, when all you want to know is "Can I fly from here to there?" But it is something you should understand.

You must know your capabilities or limitations, so cooperate with the meteorologist by telling him what they are. Being so informed, he is better able to help you. And on that subject, E. P. Norwood, meteorologist in charge of the U. S. Weather Bureau station at the Oakland International Airport, California, will now take over the discussion. . . .

When you have told the meteorologist your destination and your limitations, he will study his maps, charts and sequence reports and then tell you what weather you may expect along your route. He will say whether and where you will encounter clear weather, where scattered clouds, broken clouds or overcast. He will give details on the heights of cloud bases and possibly their tops; restrictions to visibility, if any; winds aloft, giving their direction and speed at almost any altitude you may want to fly; areas where turbulence may be expected. The meteorologist may even suggest an alternate route where weather may be better, or the upper winds more favorable.

Meteorologists may not always be right, but heeding their advice has saved many lives.

Sequence Reports

A sequence report is a report of weather conditions as observed or measured at a weather observing station and transmitted in a kind of meteorological code via teletype circuit.

One station's report follows another's in a pre-determined sequence.

The report is made up of symbols, numerals, abbreviations and a few words (see Teletype Sequences illustration) and transmitted each hour. In this way a great deal of information can be passed along in only one short "sentence."

For example, a sequence report looks like this:

```
ICT 12⊕15 ⊘85/72/55↑2⊘+29/982 QAGOM QALOM
```

Like radio stations, weather observing stations are designated by call letters. In our example, ICT is Wichita, Kansas. The next grouping, 12ϕ⊕15, tells us two things, sky coverage and visibility . . . 12,000 feet (12ϕ) scattered (⊕) and 15 miles visibility.

A report may contain more than one sky condition symbol except when the clear symbol is used. The symbols for describing sky conditions are:

○	Clear	No clouds or clouds cover less than 1/10th of the sky
⊕	Scattered clouds	1/10 to 5/10 of the sky is obscured by clouds
⊕	Broken clouds	6/10 to 9/10 of the sky is obscured by clouds
⊕	Overcast	More than 9/10 of the sky is obscured by clouds
x	Obscuration	Sky not visible at all because of some obscuration that extends upward from the ground (fog, dust, snow, rain, etc.)

To indicate that you can see through the cloud or obscuration, a minus sign (−) may precede the appropriate symbol.

The 15-mile visibility indicates that the observer could see and identify something that many miles away. The statute mile is used for visibility distances.

The next group, ⊕85, is the sea level pressure in millibars. The initial "9" or "10" is omitted. The reading here is 1008.5.

The next two groups 72/55 are the temperature, 72°F., and dew point, 55°F.

The 13 indicates the wind is from the south, and the 20+29 means the velocity is 20 knots with gusts (+) to 29 knots.

The last three figures, 982, are the altimeter setting in inches of mercury (in. Hg.). The initial "2" or "3" is omitted. But of the three figures given, if the first is 5 or higher, the initial figure is assumed to be a 2; if the first is 4 or lower, initial figure is assumed to be a 3. Here 982 assumes a 2, making the setting 29.82.

The last two groups of letters, Q-code, refer to the condition of runways, fields or navigational aids. The first in this sequence means the Instrument Landing System Glide Path (AG) is out of operation for maintenance (OM). The second is the ILS Localizer (AL) is out of operation for maintenance (OM).

Supposing that in Tucson, Arizona, the observer reports: TUS 2ɸ⊕ M35⊕/⊕5R — ɸ75/62/57 1315+/962/G27 RB35 He wants us to be able to visualize exactly what he has seen. Using the "group" method again, by the 2⊕ he has told us that he has seen clouds which obscure not more than half the sky (⊕ = scattered clouds) and the base of these clouds are 20-hundred, or 2,000 feet above the ground.

Above the scattered clouds the observer sees other clouds that he is able to measure (M) as being 3500 feet above the ground. If this second layer of clouds is directly above the scattered layer, then the second layer alone obscures 6/10 to 9/10 of the sky. If the two layers of clouds are in different parts of the sky, not one above the other, then the amounts of the two layers also obscure 6/10 to 9/10 of the sky (⊕ = broken clouds). Above these two layers of clouds, the Tucson observer tells us that he can see still other clouds.

By the slant (/) he tells us that the third layer is high (20,000 feet or more) and by the overcast symbol (⊕) that nowhere, or almost nowhere, can he see any blue sky. In addition to describing the clouds, the observer also tells us that the visibility is only 5 miles (5) and that rain (R) of light (—) intensity is falling.

With that group of symbols, letters and figures totalling only eleven characters, our Tucson observer has given us a lot of information, but he has more.

More Weather Cryptography

He tells us that the sea level barometric pressure is 1007.5 millibars (⊕7), the temperature is 62°F, (62), the dew point is 57°F. (57), the wind is blowing from the southeast (15) at a speed of 15 knots (15) with gusts (+), the altimeter should be set at 29.62 (962), the wind gusts reach a speed of 27 knots (G27), and the rain began at 35 minutes past the hour (RB35).

There are a number of letters that are used for "weather" or obstructions to visibility:

T	Thunderstorm	RW	Rain showers
A	Hail	S	Snow
AP	Small hail	SG	Snow grains
E	Sleet	SP	Snow pellets
EW	Sleet showers	SW	Snow showers
IC	Ice crystals	ZL	Freezing drizzle
L	Drizzle	ZR	Freezing rain
R	Rain		

Tornado, waterspout or funnel clouds are not abbreviated.

BD	Blowing dust	H	Haze
BN	Blowing sand	D	Dust
BS	Blowing snow	F	Fog
BY	Blowing spray	GF	Ground Fog
K	Smoke	IF	Ice Fog

Degree of intensity may be indicated by a + for heavy, − for light, and −− for very light. Moderate condition has no sign.

In talking aviation weather, the term "ceiling" is often used. Just what is meant? Quoting from the weather observer's annual: "The ceiling is the height ascribed to: (1) the lowest layer of clouds or observing phenomena aloft that is reported as broken or overcast and not classified as thin; or (2) surface-based obscuring phenomena (obscuration) not classified as 'partial.'"

In our Tucson example, the observer indicated that the ceiling was measured at 3,500 feet (M35). There are letters to indicate how ceiling heights are determined.

M	Measured ceiling
A	Aircraft (reported) ceiling
R	Radar or Raob (electronic equipped balloon) ceiling
B	Balloon ceiling
W	Indefinite ceiling

D   Persistant non-cirriform ceiling
E   Estimated non-cirriform ceiling
U   Indeterminate cirriform ceiling

The "remarks" section at the end of the teletype sequence uses abbreviations and contractions for brevity. Here are a few.

PRESF—pressure falling rapidly
BINOVC—breaks in overcast
VSBY 1V2—visibility variable 1 to 2 miles
VSBY W3—visibility west 3 miles
CBN—cumulo-nimbus to the north
FEW CI—there are a few cirrus clouds
F DSIPTG—fog dissipating
CIG 15V23—ceiling variable 1,500 to 2,300 feet
NOTAM—notice to airmen, refers to published information
PIREP—pilot report
OCNL SPKL—occasional sprinkle
RB 35—rain began 35 minutes past the hour
RE 20—rain ended 20 minutes past the hour
CU FRMG—cumulus forming
Q—code—refers to condition of runways, field, or navigational aids. Key to the code is published in the FAA Manual "Communications Procedures" (QURUA, QIOES, etc.)

Area and Terminal Forecasts

Although sequence reports give a good weather picture at the time of the observation, by the time a pilot is in the air they become too old to be of much value.

After leaving the ground you want to know what the weather will be as you proceed, and what you will find when you arrive at your destination.

At the weather station there are forecasts for areas covering almost any route you may want to fly. These forecasts show what weather can be expected; where icing may be encountered; the altitude of the freezing level; and where and when turbulence will be found. The forecasts are written using symbols and abbreviations, but with a little practice they are not difficult to read.

Teletype sequences from the U.S. Weather Bureau will give you practice in deciphering the code-like information.

```
BUF /⊕8 162/62/ 49↑↗14/001
SYR E85⊕15+ 190/56/38↑↖13/008/RE35 QIJES 28 N SIDE
UCA E45⊕10RW- 226/48/42↑↖10/018   LAMANNA QOAUA
BGM 4↖⊕E80⊕15+ 217/50/32↑↗15/016 WNBF TV QIOES
ELM E100⊕12 213/54/42↑10/015
TOL /-⊕15+ 160/66/58↗12/001
FDY O10 173/69/60↗8/004


SPI /-⊕15+ 154/68/56↑↗8/999
VLA /-⊕15 173/66/59↑5/004
SGF U⊕15 149/69/56↑11/000
MLC /-⊕15 146/73/58↑↖10/999 QIBES
FSM U⊕20 159/70/60↖3/001/ QANES QENIO + SEE NOTAMS
LIT O12 172/67/61C/004


RSL O15 051/70/56↑18+25/974/ 303
SLN O15 068/72/57↑↗16+24/976 /210
TOP 10⊕⊕15 105/↑3/59↑↗15+20/986/ 215 1070
LHX O15 017/55/28↘12/973 203 24650
TAD O30 028/46/23→6/980/ 803 24648 QAROG
DDC O15 050/69/55↑20/976/ 002 24741


MLI 10⊕⊕U⊕15+ 125/70/54↑↗13/991/ MOON DMLY VSBL
STJ E120⊕15+ 74/49↑20+32/983 SEE NOTAM GS
NUU /-⊕15 114/74/57↑12+22
CBI /-⊕15 133/72/53↑9/994
STL /-⊕15 158/69/53↑5/001
PIA /-⊕15 156/64/55↑8/999/ WMBD RDO QIOES


ZUN O15+ 064/40/19↘↙7/990 703 24732
GNT O15+ 050/43/11C/990 805 24720
FMN O15+ 041/48/1↙2/982 FEW CI W 400 24781
GBN O15+ 075/64/33↗14/977 110
FHU O15+ 049/57/25→↗10+21/990/ 605 24762
DUG O15+ 054/63/26↑↗20/988 807 24792
```

The number of airports and landing strips is too great for the Weather Bureau to provide routine forecasts for all of them. However, terminal forecasts are routinely made for many airports. For special flights, the forecaster will give you a forecast for almost any airport from which he has weather information.

Terminal forecasts in written form use about the same symbols and abbreviations as found in the sequence reports. Prepared at 6-hour intervals, they show the weather (rain, snow, fog, etc.), sky cover (clear, scattered, broken, overcast), ceiling height, surface winds, and any changes in these elements that are expected.

Winds Aloft

Observations of winds aloft are made every six hours with balloons, or a combination of balloons and electronic equipment.

Balloon observations may be prevented by surface weather conditions such as rain, snow, fog or low clouds, or they may be obscured by clouds or blown out of sight before they have reached the desired altitude. On the other hand, the balloon with electronic equipment is not affected by surface weather or clouds and is seldom blown too far for good results.

On a clear day a balloon will usually provide wind directions and speeds up to 30,000 or 40,000 feet. For the balloon with electronic equipment, information to 100,000 feet is not unusual.

In written form, a winds aloft report is composed of groups of numerals of four or five digits.

```
LAX 11951   2307   21908   1810   41710   1808
    62308   2716   82824   2830   02735   22644
    42543  62645   82655  02666   32577   51581
    02590  52580   02656  52437   02533   82410
    01506
```

LAX indicates where the observation was made . . . in this case, Los Angeles.

The first group of numerals is used to show the time the observation was made, the type of equipment used, the surface wind direction and the surface wind speed. (This group is of very little value to a pilot.) The second group, composed of four digits (2307), gives the wind direction and speed at 1,000 feet. To learn the direction, add a zero to the first two digits (23) and you have the true direction in degrees (230°). The last two digits (07) give the speed in knots (7 knots).

In the next group of numerals (21908), of five digits, the first one (2) indicates the altitude of the wind (2,000 feet). The remaining four digits, like the preceding group, indicate wind direction to be from 190° at a speed of 8 knots.

The third group, again only four digits, is decoded in the same manner as the first group and gives data for 3,000 feet.

From the surface to 10,000 feet, alternate groups have only four digits and are data at the odd thousands of feet—1,000, 3,000, 5,000, etc. The five-digit groups give data for the even thousands of feet—2,000, 4,000, 6,000, etc.—with the first digit indicating altitude.

From 10,000 to 20,000 feet only the even thousand levels are given, and the height-designating digits are 2, 4, 6, 8 and 0.

Above 20,000 feet, 23,000 is given the height-designating numeral of 3. After that, altitudes at 5,000-foot intervals are reported with a "5" for heights of 25,000, 35,000, 45,000, etc., and with an "0" for heights of 30,000, 40,000, 50,000, etc.

Like sequence reports, winds aloft reports soon become obsolete since they are made only every six hours. The information may be several hours old by the time it is to be used. To help overcome this problem, *winds aloft forecasts* are available.

LAX 3-2715/7   5-2920/2   10-3125/−1   15−3130/−10
     20−3050/−19   25−2455/−29

Decoded, this winds aloft forecast would be: Los Angeles, 3,000 feet, wind from 270° at 15 knots, temperature 7°C.; at 5,000 feet, wind from 290° at 20 knots and 2°C, etc. By interpolation it is possible to determine the winds at intervening levels.

In weather reports, whether at the surface or aloft, wind directions are true, not magnetic, and wind speeds are in knots, not miles per hour.

. . . and so ends our meteorologist's lucid account of activities within a weather station.

Those Unknown Factors

Do I hear a snicker? Maybe you've undertaken some crosscountry flights when the forecast weather didn't match up with the actual weather. It has been known to happen. But, nonetheless, don't belittle the weather forecaster who works with known information to make his prognosis. Sometimes unknown factors interfere to make the forecast unrealistic.

On a particular flight from Amarillo, Texas, to Oklahoma City, my flying partner and I checked the route weather carefully. It was summer and late afternoon—an ideal combination in that geographical area for thunderstorm activity.

Peculiarly, the sky at Amarillo was absolutely clear with just a puff of cloud here and there. We were told that the anticipated thunder activity was expected to be a line of separate clouds (cumulo-nimbus) lying across our path. As long as they were separate we weren't worried. Off we went—and flew to Oklahoma City in the smoothest possible air—and without ever seeing a cloud.

However, on another occasion, flying with the same companion, we were in Blythe, California, headed for Tucson, Arizona. In a direct line this is about 300 statute miles. Weather information at Blythe indicated a squall line, which meant connected thunderstorms would be across our path. Figuring on openings along the line, we took off for Tucson. This time the weather man couldn't have been more right if he had personally placed the squall line across our route.

The long solid line of clouds was forbidding, but there were bright spots at intervals. We picked out the biggest bright spot within our range of vision, and headed for it. Everything was smooth and calm—until we were just about clear of the cloud that was now over us. Suddenly, we felt a jolt that yanked us by our bootstraps. We flew up in our seats as far as the seat belts allowed. The suitcases hit the shelf above them. From the shelf, sundries hit the ceiling and clattered back to their resting places.

The throttle was pulled back to slow the plane and ease some of the stress that more jolts would put on the little craft. The rate-of-climb indicator showed 2,000 feet up one minute—2,000 feet down the next. The rough ride ended after three or four lesser jolts, and then we were through the squall line and back in sunshine and smooth air. That was our first such experience, and since that time the throttle has been eased back upon entering such a situation.

Just for fun, and to refute the foregoing, weather is inconsistent frequently, not following the known characteristics.

In Trouble? Ease Back That Throttle

In what seemed to be the same situation as described above, we were flying from Jackson, Mississippi, to Montgomery, Alabama. We had already been delayed two days at Jackson because of grim weather. When the elements let up enough for us to go on our way, we were eager to do so. We had already heard that another plane had departed for Montgomery an hour or two earlier and had to fly some 40 miles off course to get around thunderstorms on the way. We were prepared for any detour as long as we would arrive safely at Montgomery.

Following the line drawn on the map, we passed over Meridian, Mississippi, while flying on top of broken low-hanging strato-cumulus clouds that leveled off at 3,000 feet. As we continued on our way, leaving the white blanket behind us, we could begin to see the high build-up of towering clouds ahead and across our path into Montgomery. As we drew nearer to the line of thunder clouds, we realized the Montgomery airport was barely on the other side of the mountainous white wall.

Just about the time we were giving serious thought to being forced to alter course to find a way through, the way opened before us. The nose of the plane was pointed down at the bright spot ahead that would lead us into the airport. Light rain fell on the plane as we made our way into the large tunnel-like opening. Actually, it was more like entering the small end of a funnel; once inside the opening, the bottom of the cloud curved up, and though it was a black ceiling overhead, the rain-filtered view before us was in sunshine.

Anticipating our jolting first experience again while flying through a squall line, the throttle was eased back slightly. We waited and waited. Before we knew it we were in the clear with nary a bounce. With the airport in sight we scooted in and parked just as "our" thunder storm arrived over the airport with its roaring voice and streaks of lightning jabbing at the ground.

Since it is of no use to throttle back after a wing has been torn off as the result of too much stress on the airplane while flying through a squall line, it is strongly recommended that *power be reduced* when anticipating or flying *in rough air* that is extremely turbulent.

Meanwhile, back in the weather office, you are shown on the chart what you can expect to find en route to your destination. Inasmuch as you are restricted to contact flight (you can see the ground) the overall picture is of great importance, especially when there is any questionable weather condition that might affect your flight.

This could be weather you might catch up to, or weather that might catch up with you while you're sleeping peacefully in your hotel room.

Flying into Wheeling, West Virginia, on a particular occasion, the weather was CAVU (ceiling and visibility unlimited). Next intended stop was Harrisburg, Pennsylvania, to the east and across the Allegheny Mountains. It was to our utter dismay to learn that Harrisburg was closed with instrument flight conditions prevailing and at the same time a front that was behind us would undoubtedly catch up to Wheeling during the night and leave us with instrument weather the next day. With no alternative but to wait, we saw the weather forecast ring true. (One or two doubting Thomases tried to fly on, but had to return.

Wheeling enjoyed our company for two nights. The morning of the third day was as beautiful as the day we had arrived . . . all the way to our destination.

So, whether or not you always like what the weatherman tells you, it's wiser to heed his advice, or be prepared to head for an alternate destination.

The Weather Map

Station reports that you find on weather maps are represented by a group of hieroglyphics that resemble man's earliest attempts to express himself in writing. This sign language saves valuable space on the weather map and still gives all the current information (within a six-hour period).

The station report includes wind force (using Beaufort Scale), wind direction, temperature, sky coverage, visibility, dew point, type of cloud and altitude, barometric pressure at

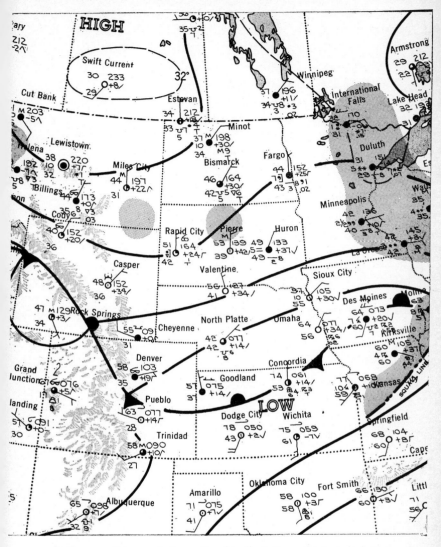

Surface weather map shows "working" station reports in addition to other weather information.

sea level, net barometric change in past three hours, precipitation in last six hours.

The Beaufort Scale assigns a number, map symbol and descriptive word to wind velocities. The map symbols not only identify strength of wind but also direction by its placement on the map. (This scale is used by all meteorologists; it is not limited to aviation weather.)

For instance, all the symbols in the Beaufort list would be indicating a west wind if they were on a weather map in those same positions. If you turn the tails straight up, the wind would be from the north. Reverse the present positions and the wind is from the east, and so on.

Isobars

Something else that you will have noticed immediately on looking at a weather map are the long curving lines that seem to be parallel. These lines, called isobars, connect points of equal barometric pressure.

Barometric pressure is the weight of the atmosphere measured in pounds per square inch, millibars and inches of mercury. Sea level pressure is equivalent to 14.7 pounds per square inch, 1013.2 millibars or 29.92 inches of mercury.

Isobars are drawn on weather maps at regular intervals of pressure of 3 or 5 millibars. They resemble contour lines that outline terrain features on maps.

There are five types of pressure areas—sometimes called pressure patterns or pressure systems—which are outlined by isobars. They are:

Low—where a low center of pressure is surrounded by higher pressure.

High—where a high center is surrounded by lower pressure.

Col—a saddleback region between two highs or two lows.

Trough—an elongated area of low pressure with lowest pressure along the center trough line.

Ridge—an elongated area of high pressure with highest pressure along the center ridge line.

Frequently, the upper air pressure charts are drawn with reference to pressure levels in millibars (i.e.: 500 MB or 700 MB).

CLOUD SYMBOLS

(As Used on Weather Maps)

	LOW CLOUDS	MIDDLE CLOUDS	HIGH CLOUDS
1.	Cumulus humilis	Altostratus translucidus	Cirrus fibratus
2.	Cumulus congestus	Altostratus opacus	Cirrus spissatus
3.	Cumulonimbus calvus	Altocumulus translucidus	Cirrus spissatus cumulonimbogenitus
4.	Stratocumulus cumulogenitus	Altocumulus lenticularis	Cirrus uncinus
5.	Stratocumulus	Altocumulus translucidus undulatus	Cirrus below 45°
6.	Stratus	Altocumulus cumulonimbogenitus	Cirrus above 45°
7.	Cumulus fractus of bad weather	Altocumulus opacus Altocumulus duplicatus	Cirrostratus covering whole sky
8.	Cumulus humilis and stratocumulus; Cumulus congestus and stratocumulus	Altocumulus floccus; Altocumulus castellanus	Cirrostratus not covering whole sky
9.	Cumulonimbus capillatus	Altocumulus of a chaotic sky	

BEAUFORT WIND SCALE

Beaufort Number	Descriptive Word	Velocity (Miles Per Hour)	Specifications for Estimating Velocities
0	Calm	Less than 1	Smoke rises vertically.
1	Calm	1 to 3	Direction of wind shown by smoke drift but not by wind vanes.
2	Light	4 to 7	Wind felt on face; leaves rustle; ordinary vane moved by wind.
3	Gentle	8 to 12	Leaves and small twigs in constant motion. Wind extends light flag.
4	Moderate	13 to 18	Raises dust and loose paper; small branches are moved.
5	Fresh	19 to 24	Small trees in leaf begin to sway; crested wavelets form on inland water.
6	—	25 to 31	Large branches in motion; whistling heard in power lines; umbrellas hard to handle.
7	Strong	32 to 38	Whole trees in motion; effort to walk against wind.
8	—	39 to 46	Breaks twigs off trees; generally impedes progress.
9	Gale	47 to 54	Slight structural damage occurs as chimney pots and slate removed.
10	—	55 to 63	Trees uprooted; considerable structural damage.
11	Whole Gale	64 to 75	Rarely experienced; accompanied by widespread damage.
12	Hurricane	Above 75	

(Note: Except "calm" the descriptive terms are not used in velocity reports.)

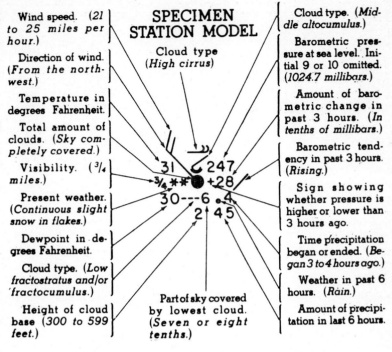

A typical station report used on the surface weather map.

The equivalent altitude in feet above sea level (MSL—mean sea level) is approximately:

Millibar Pressure Level	Approximate Altitude
1,000	400
850	5,000
700	10,000
500	18,000
300	30,000
250	40,000

(Pressure reduces as altitude increases.)

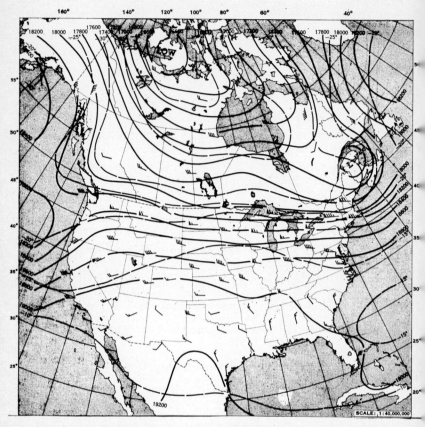

The 500-Millibar chart shows height contours that are lines (solid on the chart) of constant altitude at the 500-mb. pressure level. They indicate height above sea level. (Notice that they range from 16,600 feet to 19,000 feet.) The dashed lines are isotherms, lines of constant temperature, in degrees Centigrade. The arrows (Beaufort Scale) indicate wind velocity and direction at the 500-mb. level.

All this talk about high and low pressure has a direct bearing on your flying. Somewhere along the line you have been exposed to the subject of atmospheric pressures. You are probably saying right now, "Of course. Every time I get into an airplane I set the altimeter according to the barometric pressure."

Flying Underwater

Because the altimeter is really an aneroid barometer that is calibrated in feet, it is necessary to be conscious of the built-in errors of pressure and temperature.

With the airplane on the ground you set the altimeter according to the barometric pressure of the field's elevation. This should be the equivalent of zero altitude reading at sea level. But by the time you take off and fly to another point, the indicated altitude will almost inevitably be incorrect. You may have already experienced a landing when the altimeter showed you to be underground or maybe still in the air.

These variations are caused by different pressures along your flight route. Differences in pressures, even though at the same altitude, are caused by temperature variations and density variations of the air. Because of these reasons—although you may obtain altimeter settings from radio stations along your route—those are surface settings which will not match the actual altitude pressure.

However, before landing at a controlled airport where you can get the information, you should certainly obtain their setting on your altimeter.

Temperature affect on your altimeter while aloft may result in a reading that is above or below the actual altitude. This is caused by the air temperature not adhering to the assumed standard lapse rate of $3\frac{1}{2}°F.$ per 1000 feet. This can be caused by a layer of air below the aircraft being warmer than the assumed standard (the airplane will be higher than the altimeter indicates). Or, the reverse is true when the layer of air below the aircraft is colder than the assumed standard lapse rate (the aircraft will be higher than the indicated altitude.) This also affects the indicated airspeed in relation to true airspeed.

A rule of thumb for figuring true airspeed quickly is the addition of 3 mph per 1,000 feet of altitude to your indicated airspeed. For instance, if you're indicating 6,000 feet and 140 mph, then you can quickly multiply six by three and come up with an estimated 158 mph true airspeed.

If you want to be more accurate, you can use a computer that figures in pressure altitude and temperature. Assuming a temperature of 10°C. and figuring with the above factors on a computer, the true airspeed is 155 mph.

Pressure and Wind

There is also a direct relation between pressure and wind. Take a good look at the 500-millibar chart. Notice the isobars. Some are close together; others spread out. Each line of equal pressure is identified as to amount of pressure . . . 1020, 1016, 1004, etc. Winds generally blow parallel to the isobars, and where the isobars are close together the wind velocity will be much stronger, as the chart shows. Conversely, when they are widespread the winds will be relatively light.

The direction of the winds aloft depends on whether they are in an area of high or low pressure. On the 500-millibar chart you will see that around the Low Pressure Center the wind direction is counter-clockwise. Around the High Pressure Center, the wind is clockwise.

A change in outside temperature and pressure reading, when you are flying, is an indication of a change in wind.

Charting Fronts

Superimposed on the chart are the lines of frontal activity. These frontal lines on printed maps are represented by a line with saw-tooth (cold) shapes or half-rounds (warm) which point in the direction that the front is moving. On manuscript maps prepared in each weather station the frontal lines are represented in color—blue for a cold front, red for a warm front, purple for an occluded front, and alternating red and blue for a stationary front.

On the printed map, the occluded front is represented by alternating the cold and warm front signs along the leading edge of the line. Stationary fronts are indicated by the cold front saw-tooth sign on one side of the line and the warm front half-round sign on the other side and placed in alternating positions.

Frontal lines are drawn between air masses of different density or temperature.

Precipitation areas sometimes are indicated by green shading.

You have now completed your visit to the weather station. The meteorologist has supplied you with information concerning all the weather conditions along your route, and you're ready to go.

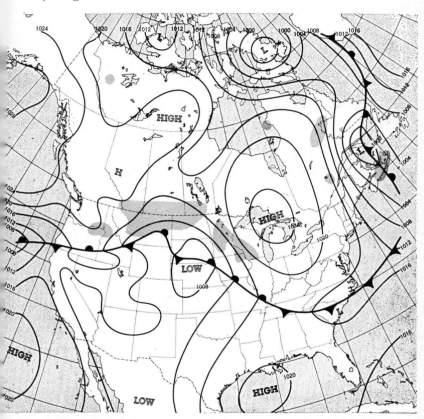

FRONTS—As seen on the weather map, fronts of all kinds—cold, warm, occluded, etc.—are illustrated by a line marked with pointed or rounded symbols to identify the type of front. This map also includes isobaric and precipitation information. The latter is indicated by the shaded areas.

3. The Fate of Fair Weather Pilots

If you claim to fly only when the weather is fair, someday you may find yourself out of CAVU (ceiling and visibility unlimited) sky with a real need to get down on the deck. Even though you are in VFR (visual flight regulations) weather you may be more than a little uneasy in the unfamiliar "weather" circumstances. If you'd prepared for the day, there'd be no problem.

Don't misunderstand. There is a great deal of VFR weather that might very well discourage the neophyte pilot from attempting a flight that could actually be safely undertaken.

To safeguard yourself from being caught in a situation you don't understand, there are two means of "preventive maintenance."

1. *Never forget the Weather Bureau is there to help you.* In the United States there is also the Flight Service Station, a facility of the Federal Aviation Agency. The personnel of both organizations are there to assist the pilot in planning the best route to his destination. The less experience the pilot has—and admits—the more detailed will be the help.

2. *Do your homework.* Once you're in the air, you are more or less on your own. This is the time that information learned about weather during your formal training period will be of help.

If you are now trying unsuccessfully to remember what you once knew, we'll try and jog your memory.

The Day of Reckoning

Once upon a time there was a young woman who had a brand new private pilot's license. She had learned to fly in the Kansas City area and was strictly a CAVU product. In the Aeronca in which she had learned, the young woman took off for Omaha, Nebraska, some 160 miles away. As she flew northward, the sky in front of her appeared to be very black and uninviting. Our new pilot hadn't the foggiest notion what was causing the threatening view.

Suddenly, there was rain on the windshield. She had never been in rain before. All she could think of was getting on the ground. The nearest airport was Shenandoah, Iowa, where she landed safely. Upset by her experience she decided to spend the night there rather than try for Omaha. The airport personnel looked at her in amazement.

"That thunderstorm will go by in a few minutes, and you'll be able to *see* Omaha from here." She wondered whether or not to believe them. But the storm did move on in a matter of minutes, and they had exaggerated only slightly. She couldn't really see Omaha, but it was clear sailing to reach her destination.

Although our young pilot was confronted by an isolated thunderstorm, it was one of some 44,000 which boil up every day throughout the world.

Thunderstorms

What makes a thunderstorm?

Simple as it may seem, a thunderstorm is nothing but a

Profile of a thunderstorm with rain is the sort of view that upset the pilot in our anecdote. As the photo shows, however, visibility is good to the left and right of the rain-obliterated view area.

cumulus shower grown up—with thunder, lightning and sometimes hail tossed in.

The thunderstorm is found in air that is highly unstable. Estimates are that every minute, day and night, some 1,800 thunderstorms roar their strength in the earth's atmosphere—discharging 100 flashes of lightning every second of every minute.

What happens is that the atmosphere tries to make the unstable air stable. In trying to do so, the unseen effort involved causes cumulus clouds to develop. They are, in fact, the visible effort of the atmosphere's attempt to regain a stable condition. The more violent this adjustment, the more severe the reaction is; the result—a cumulo-nimbus thunderstorm cloud.

Since it takes violent forces to cause a thunderstorm to form, it should be obvious that the air in or near a thunderstorm will be extremely turbulent. Inside a thunderstorm, air currents can take a plane on an elevator ride up for thousands of feet in a few seconds. That same strength can also tear an airplane to pieces. As a non-instrument pilot, you should stay well away from that situation.

Associated with thunderstorms frequently are hail and icing . . . more hazards to the aircraft. Hail can hit the leading wing edges and tear off fabric. Ice can weigh down a plane until it is out of control.

While flying around—not in— thunderstorms, beware of extreme turbulence. When it is encountered you should immediately slow your airplane's speed to whatever is recommended by the manufacturer for such conditions. At slow speed the airframe can take a rough ride in its stride.

Building the Cumulo-nimbus

The thunderstorm develops in three stages, starting with the already mentioned cumulus stage. Next is the mature stage; last, the dissipating or anvil stage.

In the cumulus stage, several such clouds may join together to form a cell. There are one or more of these cells in a thunderstorm, each behaving independently.

In this building stage, the updraft is of utmost importance because it is present throughout the entire cell. Even as the cumulus cell is building, water droplets that you might think

Early stage of a thunderhead is the flat-top cloud starting to push up. (This and the profile photo were taken just off the coast of Brazil.)

would fall as rain are carried aloft by the updrafts, or remain in suspension by the same currents.

The mature stage arrives when the suspended water droplets grow too large to be supported by the updrafts. As they start to fall, they take air with them. In so doing, they aid in forming downdrafts—the significant characteristic of the mature stage. The accompanying rain also identifies the mature stage.

These same downdrafts spread out horizontally and grow vertically. Eventually they reach the ground and cause strong and gusty surface winds. At the same time, updrafts are occurring and the result is severe turbulence. The "cell" at this point is usually 25,000 feet or more high.

The dissipating, or anvil stage, comes about when the continuously developing downdrafts and continuously decreasing updrafts result in no updrafts. This is caused by the heating and drying process produced by the downdrafts, which eventually ends the rainfall. The thunderstorm then begins to dissipate. It is during this stage that the familiar anvil top appears, while the lower portion of the cell often becomes stratiform in appearance.

The anvil stage indicates the beginning of the end—dissipation—of the thunderhead. This one was in Mexico. (Photo by J. W. Hodgkin)

How many of us assume that the anvil topped thunderhead is the cumulo-nimbus, a single thunderstorm? The fact is that this anvil top is only the indication that one cell has completed its 3-stage cycle. There very well may be other cells within the same cloud formation that are in the process of developing.

Thunderstorm research shows that most storms top off between 25,000 and 29,000 feet. Less than 10% develop to 50,000 feet or higher. The average height is 37,000 feet.

Thunderstorms are identified with cold, warm and occluded fronts (which will be discussed later) and with "air masses."

Air mass thunderstorms are of two types, convective and orographic. They both form within a moist air mass, and they usually are isolated or scattered over a wide area.

Most common is the convective thunderstorm which occurs over land or water in most areas of the world, and is the common summer storm type.

Convective cumulus clouds form over land in the afternoon when the earth is receiving maximum heat from the sun. When the air contains the necessary amount of water vapor and is sufficiently unstable, cumulus clouds will develop into thunderstorms. They also develop along coastal regions when cool

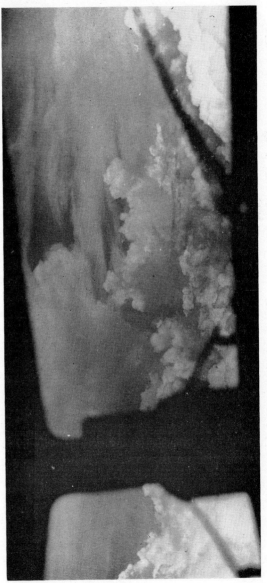

Thunderheads to the right and left as this airplane flies between cumulonimbus build-ups over one of the Philippine Islands. (Photo by Leo Cohen)

moist air from the water is heated as it moves over the warmer land surface in the afternoon. These storms will also form over water at night as cool air from land is drawn over the warmer water surface.

Orographic thunderstorms form when moist unstable air is forced up and over mountainous regions. They develop rapidly and can cover a large area. They are frequently stationary for several hours on the windward side of mountains or hills. Hail is commonly associated with these storms.

From the windward side the orographic thunderstorm is often hidden in stratiform clouds, making recognition difficult. However, from the lee, or downwind side, identification is easy —it looks like what it is.

Tornado

A storm unique to Australia and the United States is the tornado. This small but violent storm is called a waterspout when it occurs over water.

The tornado's path may vary in width from a few feet to more than a mile and in length from a few feet to nearly 300 miles.

In the United States this storm occurs generally in the midwest and along the Gulf of Mexico. However, tornadoes have been reported in nearly every state and at all hours of the day and night. June is the peak month for the storm in the midwest. Along the Gulf Coast the greatest number of tornados occur in May.

The funnel-shape storm travels an erratic path at speeds of from 15 to 58 miles per hour. But inside the storm, winds may be as high as 500 mph. Surface winds with the tornado are usually southwesterly or westerly, although they are known to have moved in all directions.

Thunder, lightning, hail and heavy rain usually accompany the tornado.

Anyone confronting the funnel-shape tornado is warned that if the tornado seems to be standing still, it's time to move . . . the tornado is headed straight for you. The best protection from any tornado is to determine which way it is going, and head the other way.

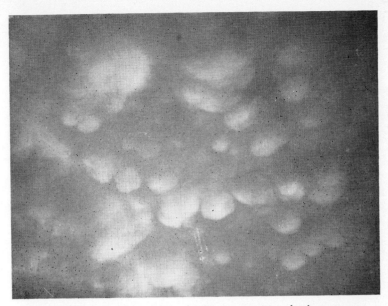

The pendulous-looking strato-cumulus-mammatus clouds are produced by strongly conflicting air masses and often herald the oncoming of tornadoes. These clouds were above the plains of mid-west United States. (Photo by Joe Christy)

Cumulus clouds build higher and higher and illustrate the situation that resulted in a landing at Fort Wayne. (Photo by Don Downie)

4. Fly for Fun and Still Reach Your Destination

The pilot reported difficulty in controlling his twin-engine airplane and requested permission to climb from his 15,000-foot altitude to 17,000 feet in an effort to get out of icing conditions. The request was granted, but shortly thereafter the pilot called in again to say that the plane couldn't get to 17,000 feet and he would, therefore, try to return to his last point of departure. That was the last ever heard from him.

On finding the wreckage two days later, it was apparent that the plane had crashed completely out of control.

This was the case of an experienced pilot flying in familiar territory. But even *he* couldn't combat the element of ice that weighted down his plane, took it out of his hands and forced him into an uncontrolled plunge to earth.

Here was a pilot with considerable flying background and training; a man who had the self-reliance to continue his flight after the first signs of icing. Without such pilot background you are not likely to find yourself in such a predicament. Nonetheless, there is always the unexpected.

A new, gung-ho type pilot making his first trans-oceanic flight for the U.S. Army Air Corps—as it was called at the time—had island-hopped across the Pacific and was on the last leg to Brisbane, Australia, from the island of New Caledonia. It was the spring season.

Up until this point the flight had been more or less routine, but on this leg they were suddenly confronted by weather. The aircraft's navigator suggested to the pilot that they leave their 8,000-foot cruising altitude and fly low under the menacing weather, even if it meant at wave-top level.

The pilot preferred having altitude between him and the ocean, although, as his navigator pointed out, it isn't any wetter going into the ocean from 100 feet than from 10,000 feet. But the pilot chose to climb, and the big plane eased its way up to a higher altitude. At 14,000 feet they had not yet topped the weather when suddenly the nose "fell off" in a stall.

Peering through the windows, the crew could see the disastrous formation of ice on the wings and control surfaces. The sudden stall of the airplane happened so fast, the navigator related, that it was not a case of slow ice build-up with an increasing sluggishness to control responses.

As the big plane nosed down, the pilot literally fought to gain control of the craft. He managed to prevent the plane from getting into a completely uncontrolled spin. The plane descended in a tight spiral. Down and down it went, but still the ice clung to the wings and control surfaces.

Slowly, as the plane dropped into warmer air and the ice began to melt, the controls became responsive to the pilot's pressure. And at nearly wave-top altitude, the plane leveled off and the terrorizing experience ended with the ultimate safe arrival at Brisbane.

Ice!

Ice comes in two forms, clear ice and rime ice. Neither type is good for airplanes. In fact, both types are bad for airplanes.

You are familiar with both types, whether or not you realize it. Rime ice is the kind that forms on refrigerator coils. It is snow-like, white and granular, easy to flake off. Clear ice makes ice cubes. It is hard, glassy and difficult to break loose.

As you know, ice forms when moisture in liquid form hits or falls below the temperature of freezing.

Since we have already learned that clouds are made up of moisture particles, it's easy to understand how freezing temperatures will result in ice formation—especially on the leading edges of aircraft wings and propellers passing through clouds or moisture-filled air at near-freezing temperatures.

The speed of the wing moving through the moisture-filled air at the near-freezing level will cool the air that comes in contact with the leading surfaces of the aircraft to the freezing point.

For instance, when a tiny drop of water strikes the leading surface, it freezes immediately in granular shape . . . rime ice.

On the other hand, a large drop of water hits the aircraft surface and spreads out before freezing in a hard thin sheet ... clear ice.

Ice in even small quantities will increase the plane's stalling speed. Ice that is on the aircraft when you come in for a landing will let you down rather heavily if you don't keep a higher-than-normal airspeed.

We've discussed ice on the leading surfaces of the aircraft, but the propeller is just as susceptible to icing as the fixed leading surface is. Because the propeller is moving at a much faster rate than the fixed wing in the ice-making air, the edges of the propeller are even more ready to build ice. As ice builds, the shape of the blade is altered and so is its efficiency. This can get so bad that the blades lose their thrust and flying speed cannot be maintained.

The propeller develops thrust from the engine. As ice increases on the propeller, more of the engine power is used to overcome the prop's decreasing efficiency. Increased drag is the result. This transfer of engine power to overcome increasing drag caused by ice can continue to the point of disaster.

Inasmuch as most light airplanes are not equipped with de-icing gear, it is best to stay out of ice-inducing conditions. If you do get caught in such a situation, lower altitudes usually offer warmer temperatures ... assuming you can get down to a lower altitude and still leave mountains and other geographical landmarks intact.

Carburetor Ice

As a private or student pilot, probably the most important form of ice to affect you was introduced with your first flying lesson ... carburetor ice. This ice differs from rime and clear ice in that it does not depend on visible moisture in the air, nor on freezing temperatures, to form.

It can form when outside temperatures are well above freezing and in clear air. Beware, however, if humidity is high.

An Invaluable Lesson

We were at 6000 feet and just about overheading Fort Wayne, Indiana, when the engine began to struggle. The weather report we had received at Springfield, Missouri, had given us the go-ahead to fly above the clouds, those puffy pretty white mounds that look almost good enough to eat with a spoon. But as we proceeded on course we had to climb higher and higher to stay above the broken cloud layer.

We took a quick peek at Fort Wayne through a sizeable break in the clouds, and proved we were on course. Our destination was Detroit, and we didn't want to waste time on the ground with an unnecessary landing.

But as the engine sputtered, we gave another thought to Fort Wayne. Of course, the first thing to do was pull on the carburetor heat. Our small plane did have the benefit of a mixture control, and it was leaned out as much as our altitude (now climbing past 9500 feet), would permit.

The addition of carburetor heat helped only briefly. The engine lugged, struggled, strained and gasped for a breath of fuel. With only partial power, and clouds rising higher ahead and suddenly looking almost solid under us, we decided to turn back for that hole over Fort Wayne.

As we reversed course, we spotted another opening in the cloud deck below us, and down we went. As soon as we hit lower altitudes our faithful little engine began to breathe freely once again, purring happily on all four cylinders. Still, we considered, even though flying low might solve our problem, perhaps we'd be better off with a check of the engine.

Back to Fort Wayne we went. And it was there that we learned an invaluable lesson on judicious use of carburetor heat. (Oh, we did have a dead left magneto, which made us feel better about landing, but there was a lesson for us all the same.) We were told that if carburetor ice begins to build up without being noticed, and suddenly you realize that the throttle is at the firewall and you're still losing power, close the throttle and apply carburetor heat. Maintain enough power to hold altitude, and once RPM's have picked up, reduce carburetor heat, or take it off entirely.

By this time you should be alert to the carburetor icing potential and catch the first diminishing notes of engine power with carburetor heat in normal fashion, applying only enough heat to clear up the first signs of icing condition. It can be an off-and-on situation, and one that you have to keep ahead of or you'll find yourself right back where you started.

Another experience involving this same problem happened on a low level flight from the east side of San Francisco Bay to Sacramento, capital city of California, some 80 miles away.

Because it is a relatively short flight over flat, nearly sea level terrain, it isn't often that one climbs very high to make the run. This particular occasion was no exception. Flying the same Taylorcraft that started this book, I was chugging along at something around 2000 feet. This was winter and there was a broken ceiling with some rain showers and a tremendous tail wind. Happy in my ignorance, I was more than a little startled when the engine started making groaning sounds.

I reacted as I had been taught and applied carburetor heat, but it took my companion (who just happened to be a U.S. Air Force pilot, *and* my brother) to complete the cure. In this case, he applied full power, then pulled it off, then re-applied it. After several push-pulls, there was a loud *Pop!* and RPM's returned to normal. We had backfired the ice out of the carburetor.

You may have already had your own experience with carburetor icing. There you were, cruising peacefully along, when you noticed the power drop off slightly. You increased the throttle setting so that the tachometer indicated your proper cruising RPM. But the same thing occurred again. You increased the throttle setting once more, only to realize that you were at full throttle and still your RPM's were slowly dropping.

We'll assume that you were saved from engine failure by adding carburetor heat, because that's what your flight instructor told you way back. But what was going on inside the carburetor to cause this power loss?

When humidity is high, you'll find a large amount of water in the carburetor. This is the result of the air vaporizing as it is taken into the carburetor to mix with the plane's fuel. As it does so, the air is cooled to freezing—or even lower. At the freezing temperature, the moisture in the air forms ice around the opening to the carburetor. As ice builds up by the continuous rush of air, the amount of air permitted into the carburetor diminishes.

If the process is permitted to continue, things only get worse.

With diminishing air in the carburetor, but the same amount of fuel, a rich mixture results—and the engine begins to lose power. (You know yourself that as you go through the pre-take-off check list procedure, when carburetor heat is applied, the RPM's drop. They are supposed to because the outside cold air is cut off.)

With the engine losing power, the pilot increases the throttle setting—which increases the flow of fuel into the carburetor, where less and less air is entering. The result, a richer-than-ever mixture with power decreasing at a greater-than-ever rate. Eventually, the carburetor is completely shut off to additional air, and the end result is no power at all.

Maybe you're saying to yourself that you did get into such a situation, but you knew enough to lean out your mixture (not all light planes are so equipped) so that the limited air intake had less fuel to mix with, thereby resulting in normal fuel-air ratio. If so, good for you!

But leaning the mixture isn't the sole solution to the problem. You, apparently, flew out of the air conditions that caused carburetor ice before you could get into further trouble. Because even though you lean the mixture, eventually—conditions being the same—the carburetor ice will build up so that even your "lean" setting will become "rich." If permitted to continue, the engine will suffer from the same fuel starvation that the plane did whose pilot continued to increase the throttle setting.

Now that you're wondering just how to escape the carburetor icing problem when the two obvious paths seem to be closed to you, read on. The answer is simply this: anticipate icing before it occurs. At the very first sign of power loss, or a rough engine, apply carburetor heat. But don't leave it "on." By doing that, after you have disposed of the icing problem, you will have another problem in loss of engine power. A warm mixture means a greater expansion of fuel before it is compressed in the cylinders. This results in a lower compression ratio and, consequently, lower engine power output.

For this same reason, make sure the carburetor heat is "off" —at time when you want maximum power.

If carburetor ice has built up so that just adding heat does not solve the problem, there are the "last resort" methods related in the foregoing experience stories. To repeat, one method is this: with carburetor heat on, reduce power/throttle setting to minimum and still maintain altitude. This will mean a nose-high attitude. The reduced onrush of moist air into the carburetor will minimize the icing. After a minute or so, return to normal cruise and throttle setting.

The other method is to reduce power setting with carburetor heat on. Push and pull the throttle in and out with the hope that the engine will backfire, thereby loosening the ice and opening the carburetor air intake valve. These are not guaranteed, but they have worked successfully.

But the best way out is still to anticipate icing conditions, and be prepared to "play it by ear" with the use of carburetor heat to control ice.

Ice Dangers on the Ground

Ice on the ground is also a hazard to be reckoned with. Frost, sleet, frozen rain or snow can accumulate on parked aircraft, and must be removed before starting the engine. If there is water or wet mud on the ground (parking ramp, taxiways, runways) the propeller can blast this onto the leading edges of the wings, where it can freeze and be dangerous.

It is considerably wiser to take the easy way out if you find yourself headed for trouble, or even if you only think you're headed for trouble. It is the better part of discretion to be a live coward than a statistic. The braggart who tells how he made it through a tight situation that he should never have attempted is not only boring to other pilots but is also a likely prospect not to make it through the next one.

Atmospheric Pressure and Wind

It has been mentioned on a preceding page that when there is a change in pressure and temperature, there is most likely to be a change in wind. The three are interrelated. We'll go into this a bit further here.

As you remember, atmospheric pressure is the force exerted by the weight of the atmosphere on a unit area from the level of measurement to its outer limits. In other words, imagine an endless square-foot column of air perpendicular to the earth's surface. At the surface the atmospheric pressure from that column of air is 2,116 pounds per-square-foot, or 14.7 pounds per-square-inch. Each foot up the column the pressure decreases, but it does not decrease at a regular rate. This is due to the fact that variations in air density which are caused by variations in the distribution of temperature result in variations of pressure.

Over land areas heated to a high degree by the sun, there is usually a decrease in pressure. But over the adjacent areas there is a *rise* in pressure. Why? As the hot air rises it eventually tends to spread out. By doing so, the hot air reduces its own pressure in the space where it has risen, and as it spreads it is forcing more air into the adjacent areas—thereby increasing the pressure in those areas.

When cold air is drawn (advected) to another area, the weight of the air is increased, thus increasing atmospheric pressure. On the other hand, when warmer air is advected to another region, pressure decreases.

Because pressure variations are important in relation to wind, clouds and fog, much attention is given to both surface and upper air pressure and the differences noted. These pressure readings are made at weather stations at the same times, then marked on special maps for study and comparison. The lines connecting points of equal pressure (isobars) are drawn.

Repeating what was said earlier, wind will tend to blow in a direction parallel to isobars. Where the isobars are close together, winds will be strong; where they are far apart, winds will be light. The wind's velocity is, therefore, directly proportional to the steepness of the pressure gradient or height. The wind direction will be dependent on whether it is in a low or high pressure area. Around a low pressure area, winds blow counter-clockwise. Other general patterns of wind circulation will be discussed in the next chapter.

The pilot who admits he lucked through an experience that he'll never try again is much more likely to be the pilot described in the old familiar saying:

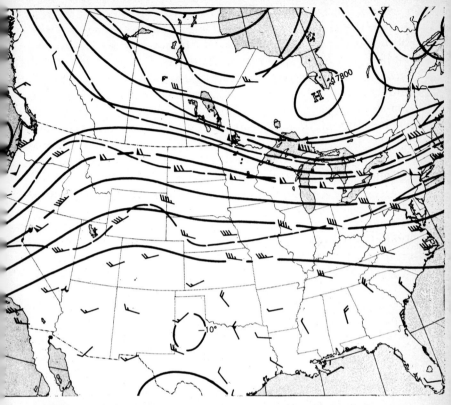

Isobars and wind direction are more or less parallel as this map indicates. These isobars (solid lines) are at approximately the 18,000-foot level. Notice that the closer together the isobars are, the greater the wind velocity is.

"There are bold pilots and there are old pilots
But there are no old, bold pilots."

Flying is for fun and utility. When weather is unfamiliar to you and marginal for your capabilities, sit it out.

If you're on a time schedule that requires you to be on your way, there is only one safe way out, miserably annoying though it may be—take a bus. It may be a blow to your ego, and a nuisance to retrieve the plane at a later time, but it's better to be alive to be able to gripe about the experience.

High Altitude Operation

If you're basically a sea level pilot, the first time you head for a high altitude airport you will have some new experiences awaiting you. The air is less dense than at sea level. This has an important effect on your take-off at high altitudes. The most effective way of demonstrating this is by looking at the chart "Altitude-Temperature Effects."

There is nothing dangerous about high altitude landings and take-offs just as long as you remember that the take-off run will be longer. In fact, depending on your load, temperature and altitude, the take-off run can be much, much longer than you might expect. The point is this—do expect the longer run, plan accordingly, and you'll have no problem.

On the chart "Altitude-Temperature Effects," take a good look at the example shown by the line. Notice the temperature on the left scale is 100°F. Follow the line to the right-hand scale where it points to 6,000 feet altitude, or field elevation. Where that line crossed the short scale pointed to by three arrows, focus your attention carefully. On the scale at the right, the per cent of decrease in the aircraft rate of climb is a little better than 75%. On the left scale, the per cent of increase in take-off distance is 220%!

Take a straight-edge, and make some comparisons of your own. You'll notice that the lower the temperature, the less extreme the take-off distance and rate of climb at high field elevations. For this reason it is recommended procedure to take off in the early morning or late afternoon from such airports and avoid high mid-day temperatures.

To illustrate the point, a pair of pilots had landed at Cheyenne, Wyoming, one hot summer day. Cheyenne has nearly a mile-high airport elevation. They were flying a small, 85 hp. airplane. Their departure the following day seemed routine; they had given no thought to the high altitude. They were obviously unacquainted with the Koch chart. As they taxied toward the runway, they asked for permission to take off from the mid point of the long runway. The immediate reply was a clear, concise listing of the airport elevation, the temperature and the dew point, and the final comment, "You may take off from the mid point at your discretion." The tone

THE KOCH CHART FOR ALTITUDE AND TEMPERATURE EFFECTS

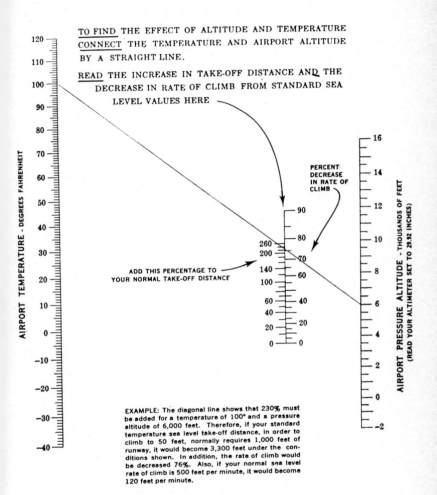

This chart indicates typical representative values for "personal" airplanes. For exact values consult your airplane flight manual.

The chart may be conservative for airplanes with supercharged engines.

Also remember that long grass, sand, mud or deep snow can easily double your take-off distance.

of the voice was enough to make the two pilots wish they'd never made such a ridiculous request, and they taxied back to the end of the runway.

This is another lesson you must remember if you want to fly for fun and still reach your destination.

5. Clouds and Where They Come From

There is cause for wonder, perhaps, when you have been soaking up sunshine on a beach under a cloudless sky and suddenly see that small scattered puffs of white clouds have formed. After awhile there are more and bigger puffs that slowly move over the water toward shore.

Something like that can also happen when you're enjoying the clean air of the mountains. In the distance you can see just the slightest bit of white cloud peeking out from behind the mountains. Slowly that cloud grows in height and size and is joined by others. Before long, you have towering cumulus monsters blasting out lightning and thunder. What, you may ask yourself, is going on?

To begin with, clouds are formed by the condensation of water vapor. An additional requirement is the presence of microscopic particles called "condensation nuclei" in the atmosphere. These particles are made up of salt for the most part. The atmosphere always has enough of the condensation nuclei present for the cloud condensation droplets to form. It happens when the air temperature falls below its original dew point at a given level.

Clouds have also been described as a direct expression of the physical processes which are taking place in the atmosphere. They are formed, just as rain, snow and other types of precipitation, when the air is cooled below the dew point at a given altitude. (Dew point is the temperature at which water droplets in the atmosphere become visible.)

The term dew point is one of two that describe the amount of water vapor in the air. However, dew point is a temperature while the other term, relative humidity, is the ratio of the water vapor actually present in the air to the maximum amount of water vapor the air can hold at a given altitude and

On the beach

In the mountains

temperature. Relative humidity is spoken of in percentages of the atmosphere's saturation point. That is, when the air contains all the water vapor possible at a particular temperature and altitude, the relative humidity is 100%.

One step beyond the dew point brings us to condensation ... the process by which the water vapor in the atmosphere is changed into visible moisture as fog or clouds.

Clouds Classified

Clouds are classified according to form or appearance and by the physical processes that produce them. Also, there is a general relation between cloud forms and their heights above the ground. There are two general categories for all clouds—*cumuliform* and *stratiform.*

Cumuliform clouds are formed by rising currents in unstable air, while stratiform clouds are formed by cooling air in stable layers. Puffy—like tufts of cotton, and usually with flat bases—cumulus clouds nearly always are distinctly separate.

The significance here is that the cumulus cloud itself indicates a rising current, while the space between clouds indicates a downward current. From this you can expect turbulence—the taller the cloud the greater the possible turbulence. Precipitation from cumuliform clouds is generally in rain showers.

Stratiform clouds are in layers without much vertical development. They can cover the sky or be patchy. Pecipitation is drizzle, light continuous rain or snow.

Divided into categories, clouds are Low, Middle or High, plus clouds with vertical development. The break-down, with abbreviations, follows:

LOW—stratus, strato-cumulus, cumulus, cumulo-nimbus
         St   Sc   Cu   Cb

These clouds are found usually below 6,000 feet.

LOW: Stratus or "High Fog" over Santos, Brazil

Stratocumulus cumulogenitus over San Francisco Bay

Stratocumulus from the top over Houston, Texas

Cumulus humilis at Honolulu International Airport, Hawaii

Fracto cumulus

Cumulus congestus and altostratus in the Philippines (Photo by Leo Cohen)

Cumulonimbus calvus along Argentine coast

MIDDLE—alto-cumulus, alto-stratus, nimbo-stratus
       Ac   As   Ns
These clouds usually are between 6,000 and 20,000 feet.

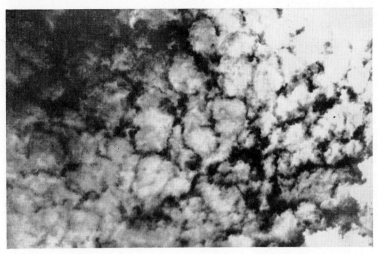

MIDDLE: Altocumulus translucidus—"Looking Up"

Altocumulus cumulogenitus

Altocumulus over Germany—"Looking Down" (This is not the recommended method of viewing clouds for the average pilot.)

Altocumulus and cirrus

Altostratus opacus (Photo by Leo Cohen)

HIGH—cirrus, cirrocumulus, cirrostratus
    Ci   Cc   Cs
Usually found above 20,000 feet and as high as 50,000 feet, they may also occur at much lower altitudes in arctic regions.

HIGH: Cirrus uncinus above 45° showing up-turned "Mares' Tails" over Texas

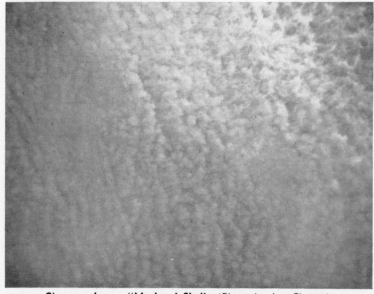

Cirrocumulus or "Mackerel Sky" (Photo by Joe Christy)

Cirrostratus with jet vapor trail indicating the high altitude of this cloud formation

The low-level stratus clouds look like fog, but they do not rest on the ground. They are usually formed in stable air which is associated with a temperature inversion. Their thickness is from a few hundred feet to several thousand. Visibility is usually poor under stratus clouds which have light rain or snow forms of precipitation.

Ragged or broken clouds have the word fracto (meaning broken) attached, such as in "fracto-stratus."

When the word nimbo is added you can expect rain, for that's what it means . . . nimbo-stratus, cumulo-nimbus, for instance.

The nimbo-stratus cloud is formed in a layer of dark gray. It is of uniform appearance and seems to be dimly illuminated from the inside. Snow or continuous rain is the usual type of precipitation. This cloud often rises above 15,000 feet.

Strato-cumulus clouds look like their name sounds—a low layer with round blobs or long rolls and waves. The wave appearance is particularly true in winter. These clouds usually form *below* a temperature inversion.

Clouds with vertical development include cumulus of all types. They have the distinctive shape of a bell, flat on the bottom, high and rounded at the top. They are thick with the whipped cream look.

Cumulo-nimbus is the superlative of the cumulus. It will grow in height before your very eyes. Flat on the bottom, it is dark underneath—where it may be raining. It builds up in billowing shapes, and is frequently topped off with an anvil formation. Its mountainous appearance is frequently added to by a flanking of lower billowing cloud masses.

This is the cloud that warns *"Beware."* It can produce rain, hail and snow—all accompanied by lightning, thunder and extreme turbulence. This is a cloud to fly around when you find it.

When it is part of a series of cumulo-nimbus in a squall line, fly over the saddles between the towering build-ups. If that is not possible, head for the light areas under the line, flying closer to the ground than to the cloud base, throttle back and prepare for a rough—if short—crossing.

Middle clouds appear in the middle latitudes of 30° to 60°. It is noteworthy that they are found nearer the 6,000-foot level in cold seasons and nearer the high level of 20,000 feet during the warm months. The middle clouds are composed of water droplets for the most part, and some forms may contain ice crystals.

Alto-stratus is a thick cloud layer that has a gray or bluish veil-like appearance through which the sun may be vaguely distinguished. Rain or snow often falls from these clouds which can lose the solid look when broken by wind.

Alto-cumulus clouds form in a layer of small broken puffy mounds. They can be in groups, lines or waves, and there can be more than one layer at a time. Alto-cumulus and alto-stratus clouds often merge at the same levels when the air stability is changing. This cloud is often confused with cirro-cumulus. About the most noticeable difference between the two is that the cirro-cumulus will be completely white, while the alto-cumulus will have gray shading effects.

Mots familiar of the cirrus are the high-in-the-sky filmy, gossamer-like clouds. Bright white, composed of ice crystals, cirrus

Cumulus Series showing a variety of the low-level puffy mounds and their flat-bottom characteristic

formations are actually quite varied. They can be isolated tufts, plume-like, or streaks with turned-up ends known as "mares' tails." The sun shines brightly through cirrus clouds.

Cirro-cumulus clouds are distinctive small puffs of white that give a ripple, seersucker effect that is referred to as a "mackerel sky."

Some Unexpected Experiences

Sometimes fliers don't know what a cloud is until they're in one. On a flight from Bakersfield, at the southern end of California's San Joaquin Valley, heading south for the Los Angeles area, a pilot started the climb to top the mountain ridge and high desert plateau that geographically separate northern and southern California. A layer of stratus clouds, as reported, was visible. At the weather station indications were, however, that the pilot could safely fly under the clouds and above the rising terrain. He started up a canyon that ended at the lowest point of the mountain profile in front of him. The cloud layer began to appear lower than anticipated, but still our pilot thought he could make out the bottom of the clouds and the top of the steep slope of mountain ahead. He had just begun to reconsider this decision when the lowering clouds and his path of flight joined.

Startled and scared, he lowered the nose of the plane and turned abruptly for a 180-degree escape. As a result, the pilot's passenger, his wife, now takes an even dimmer view of light plane flying than she did previously.

The weather report at Mazatlan, Mexico, didn't mean a thing. Low stratus was already forming at the city—which is on the Pacific Ocean coast just opposite the tip of the Baja California Peninsula. Our destination was 200 miles south, Puerta Vallarta. Off the ground and on the way, the stratus spread further and further inland. The only saving grace was the fact that "inland" ended with nearby mountains. As the low stratus became solid under us, everyone in the plane kept an anxious eye toward the mountains. I figured that if the cloud deck below didn't offer an opening at the "point of no return," I would turn around and head back to Mazatlan . . . which I hoped would still be open. About the time that mouths were getting dry, a large opening in the stratus appeared, and down we went to follow the shoreline at less than 1,500 feet for some 50 miles to our destination. There were practically no clouds in the air when we arrived, but we had no way of being sure of that before or during the flight.

And there was another time when your author took off at night from San Diego—which is almost at the Mexican border—for the hour's flight north to Los Angeles. Having been told that there were scattered clouds at 2,000 feet, it was intended to stay under that altitude since it is flat country along the coast. Whatever the excuse (it was the last leg on a flight from the interior of Mexico), the first thing that this pilot knew was that the airplane was in a cloud.

The first thing to do is panic, but don't let your passengers know it. Demonstrating sheer stupidity, when the simplest and most logical thing to do was merely point the nose of the plane down, for some reason a 180-degree descending turn came to mind. If it hadn't been that we were in the bottom of the cloud to begin with, the famous death spiral could have been the end result. Fortunately, after about a 270-degree out-of-control turn, we were back on the well illuminated highway which we followed right in to Los Angeles.

Why You Should Fear the Elements

And then there are experienced pilots who are limited to VFR flying by the aircraft and its equipment. A good example of a flight which would have been more easily undertaken in an aircraft equipped for instrument flight and de-icers started south from Seattle, Washington, for Klamath Falls, Oregon.

The pilot reported that after take-off he hit rain that nearly obliterated his vision and caused such static on the radio that it was useless for several minutes. High winds and turbulence dogged the plane to Portland, Oregon, where the pilot headed up the Columbia River Gorge to Hood River. There he landed for the night. During the flight through the gorge, he had been forced down to 600 feet by rain and clouds.

The following day the weather was a little better until he approached Klamath Falls. There he was confronted by a black wall of snow-bearing clouds that forced him to retrace his path to a small cleared strip he had noticed in the pine tree

Snow-bearing clouds surround the airport at Reno, Nevada. This was to have been a gala mass fly-in weekend, but the forbidding clouds reduced the number of participants—some came by airline—but not the fun. This was just such weather that the pilot in the anecdote gambled four lives on and lost. (Photo by Vic Stark)

forest. He landed amidst scattered large snow flakes. An open hangar offered refuge for the plane, and a nearby lodge offered refuge to the pilot.

A couple of hours later, the pilot took off again—only to be confronted by more towering snow clouds. This time he decided to climb over the wall. At 12,000 feet it was murky, but he was able to clear the clouds and make radio contact with Klamath Falls. He was told that there were large holes over the area, so when one appeared he went down. Unfortunately, he related, there was still another black cloud facing him, but by flying low he managed, with a minimum of visibility, to reach his destination.

He summed up his flight by saying that it is just such "stuff as this that can kill private pilots, or old flying dogs, too, for that matter. I don't know about others, but I personally have a lot of fear of the elements." So says a professional pilot of 30 years' experience.

A less successful ending was the fate of another pilot.

A flying school had rented a four-place plane to a former student they had trained a few years previous. The pilot had three passengers, including his teen-age nephew. After a pleasant sojourn at Reno, Nevada, they prepared for the less-than-two-hours return flight to the San Francsico Bay area.

Weather was dubious, but the pilot was confident. However, he did telephone the plane's owner. He was told not to fly if there was any doubt whatsoever. "Go home commercially, or stay there, but don't fly if it looks at all marginal."

Besides the four lives that were snuffed out by the snow storm, a brand new airplane was washed out . . . uselessly. *The moral:* never, never fly beyond your capabilities.

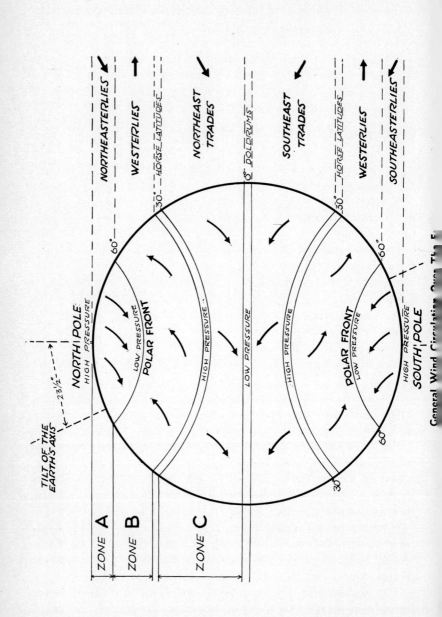

General Wind Circulation Over The Earth

6. Circulation of Air Masses

If clouds are an indication of the kind of weather which is to follow, then let's back up to find out what causes the weather that shapes the clouds.

A general flow of air of the atmosphere is found around the earth. In the northern hemisphere there is a certain deflection of the air movement at high levels toward the North Pole, and at low levels toward the equator. This is reversed in the southern hemisphere.

Refer to the chart "General Wind Circulation over the Earth."

Around the equator the air is warm. This gives it a tendency to rise, causing low pressure.

The cold air of the polar region is heavier and denser, resulting in high pressure. It tends to sink and flow towards the equator.

You've heard of the "doldrums" and the "horse latitudes." They really do exist. By any other name, the doldrums is a belt around the earth at the equator.

From the equator north to 30° is the area of the Northeast trade winds (Zone C on the chart). Next are the horse latitudes, a belt at 30° latitude. Prevailing westerly winds are between the horse latitudes and the polar front at 60° (Zone B). Then follows the polar region (Zone A) from 60° to the North Pole.

In the area of the doldrums, light and variable winds from both northern and southern hemispheres converge. This is also an area of heavy rainfall, frequent showers and thunderstorms.

The region from the equator to the north side of the horse latitudes is marked by low humidity and generally clear skies. Lack of rainfall here makes it home to most of the world's deserts.

Since winds tend to blow from high to low pressure areas, the prevailing winds flow south from the horse latitudes to the doldrums (Zone C). These winds moving south are deflected to their right; hence, are known as the Northeast Trades.

The polar region is bordered by a low pressure belt to the north called the polar front, and the high pressure of the horse latitudes to the South (Zone B). Winds here, stormy and extremely variable, are westerly since they flow from a high pressure toward a low pressure area and are deflected to their right as westerly winds.

North of the polar front (Zone A), with low pressure to the south and high pressure at the North Pole, winds are northeasterly.

In the southern hemisphere, wind directions are southerly but otherwise circulation is much the same in both hemispheres, as the chart illustrates.

There are, of course, local patterns of circulation resulting from land, water, mountain and thermal effects in both hemispheres.

With this general picture in mind of the patterns of air movement in the northern hemisphere, let's go on to the air masses that are involved in these general air movements in the atmosphere.

Air Mass Movements

The original characteristics acquired by an air mass when it is developing identify the source region. There are only two possibilities— tropical or polar; and for further identification of the air mass properties, either continental or maritime.

When tropical air (warm) meets polar air (cold), they do not mix. When the two come together, a line on the ground is formed and is called a "front." More about fronts later.

Getting back to air masses, polar air (cold) is so called because it is colder than the earth's surface over which it is passing. At the same time, however, the cold air is warmed by the earth and vertical air currents are formed. These are the cause of "bumpy" flights.

Looking through a vertical build-up of clouds of the cumulonimbus type. These clouds are along a shoreline where they have picked up additional moisture and are producing rain as the result. (Photo by Leo Cohen)

Looking under a cloud of vertical build-up such as described in the preceding photo, one can see rain falling from this over-water cloud.

Not only will air currents be vertical, but clouds will also have vertical development—such as cumulus type clouds. If this cold air mass is moving over land, fair weather may be anticipated because little moisture is picked up. Conversely, however, if the cold air mass is moving over water, more moisture is picked up and carried to higher levels, where cumulonimbus clouds may form and produce rain.

Good surface visibility may be expected from a cold air mass because the vertical currents in the cold air pick up smoke, haze and moisture and carry them aloft.

A cold air mass is favorable to the pilot because he can expect good ceiling and visibility near the surface. It will be recognized by a bumpy ride and cumuliform clouds.

What about the warm air mass? It is identified as any mass of air that is warmer than the earth's surface over which it is passing. Ocean areas in the horse latitudes are the most important source of warm air masses.

The air is warm, fairly stable and high in moisture content. When it moves to colder regions the lower layer cools and becomes even more stable. Cooling prevents turbulence and shuts off any vertical currents, leaving the air movement almost entirely horizontal.

For the pilot the warm air mass offers smooth air. At the same time, however, visibility near the surface is reduced, and ceilings tend to be low, with cloud forms of the stratiform or sheet type.

Turbulence

The amount and intensity of turbulence depends to a great extent on the instability of the air. That is, the less stable, the more turbulence . . . and rougher, bumpier flying.

Turbulence is attributed to a variety of causes . . . convective currents (thermal or heat), irregular terrain (mechanical), lifting of warm air (frontal), marked change in wind speed with height (wind shear), aircraft wake (man-made).

Even obstructions on the airport can cause turbulence or gustiness at ground level and can be dangerous. In approaching severe weather, where high winds are anticipated, it is the wise aircraft owner who checks to see that his plane is secure, whether tied down or in a hangar.

On one occasion a combination of convective currents and irregular terrain garnered enough strength to give an airplane a jolt that knocked open the ash tray, sending a cloud of ashes through the small cabin, and popped open both doors. The two pilots of the rented plane, non-smokers, hadn't thought to check the ash tray contents before departure, and their immediate thought was that the airplane was disintegrating.

Aonther experience, this time dealing with man-made turbulence, took place at low altitude over flat terrain. A four-engine airliner crossed the light plane's flight path at a right angle. The big plane was at least one mile distant when the smaller plane hit its wake. Just one positive jolt that caused one door —that was locked—to open, and the air smoothed out. The pilots in this case had to think twice before they realized the distant airliner was the undisputed cause.

(To shut doors that pop open in flight, slow the aircraft to near zero speed. There's less resistance to your efforts. It is also easier if there is someone other than the pilot to do the shutting.)

Landing Behind the Big Ones

If you are in a landing pattern and find yourself assigned to land behind an airline-size plane, be wary of wake turbulence. Put as much distance and time between yourself and the big plane as possible. If you are not able to delay your landing, it is recommended that a steeper approach angle and higher airspeed be used than normally. If there is any cross wind, land on the up-wind side of the runway. (This holds true for take-off, too.) The higher airspeed on approach will give you better control in the event you should run into wake turbulence.

Warm ground heating cooler air above it adds up to rough air. Such convective tubulence will have a height of several thousand feet. The strength of the turbulence is exaggerated over barren land such as desert. Where there is vegetation, convective turbulence is less severe.

Some Thermal Turbulence

An example of turbulence caused by thermal (or convective) activity occured during a visit to Desert Air Park at Palm Springs, the vacation mecca in the California desert 100 miles east of Los Angeles.

The month was May, still relatively cool in the desert, but the weather was not being seasonal even when we arrived.

Blowing sand and dust near Palm Springs can be seen silhouetted against the hills in the background. The grass runway helped to keep blowing dust to a minimum where planes on the ground were concerned. The small cumulus clouds indicated that the air was turbulent . . . and it was.

Winds were high with gusts blasting higher to an estimated 35 mph. And it seemed as if the desert had been picked up and was being rearranged, at least as far as sand and dust were concerned. The blowing earth didn't particularly restrict visibility, but it was a nuisance.

One place where the dust could have caused trouble was in the carburetor. With the valve left open in routine position, blowing dust could have clogged the carburetor. However, once carburetor heat was turned "on," prior to landing, it stayed in that position during the time the plane was parked up to the point of starting the take-off roll on departure.

When time for departure did arrive, the unpleasant wind-with-gust condition still prevailed. We were warned that our route north would put us in extremely turbulent air. Our flight path was over desert for 150 miles. The warning was good—up to a point.

As we climbed, the rising terrain below became hilly and craggy—ideal for adding to the turbulence that was already severe from heat radiating from the desert. But by the time we had left the low desert floor behind and had crossed the hills to the higher desert floor at close to 3,000 feet, the cooler air dissipated the turbulence, and it was smooth sailing.

Those Bumpy Summers

Hot summer afternoons are almost certain to be accompanied by turbulence. When possible, cross-country flying should be done in the early part of the day when the cool morning air provides much smoother operating weather.

Another way to avoid a rough flight is to fly above the clouds. Cumulus clouds will form in convective, or thermal, currents when the air is moist.

Turbulence caused by mechanical activities vary in intensity depending on wind velocities. They occur when the air near the earth's surface flows over hills, mountains or buildings. The disrupted wind flow is broken into a variety of eddies and odd currents.

Strong winds hitting a mountainside can result in severe turbulence in updrafts and downdrafts. On the upwind side of the mountain an airplane will get a "free ride" up. But on the downwind, or lee, side, it is possible to get the same ride in reverse—and this can be tricky.

A phenomenon known as "mountain wave" occurs when the air currents on the upwind side are extremely strong, and the air on the downwind side is extremely stable and stationary. The rising air currents on the upwind side move to the lee side as down currents that continue as far as 5 to 10 miles from the mountain top and then start to rise again in a wave. Wave action has been known to repeat itself more than six times. There is decreasing intensity with each repetition. The cloud peculiar to this mountain wave is lenticular, known to glider fans for its excellent lift-giving qualities.

Frontal turbulence results when cold and warm fronts meet.

Mountain Wave effect is illustrated here with clouds following along the downwind side of this ridge near Owens Lake, California. Characteristic clouds identify the wave. At top, the lenticular clouds; below it the roll cloud. Better defined lenticular cloud is seen in the small photograph on next page. (Photo by Don Downie)

Abrupt changes in wind velocities or directions at adjacent altitudes cause turbulence produced by wind shear (the change in wind velocity or direction). In the vicinity of jet streams, for instance, such turbulence is found. Because this happens frequently in cloudless skies, it is often referred to as "clear air turbulence."

Lenticular clouds usually form around 20,000 feet or higher. (Photo by D. L. Morehead)

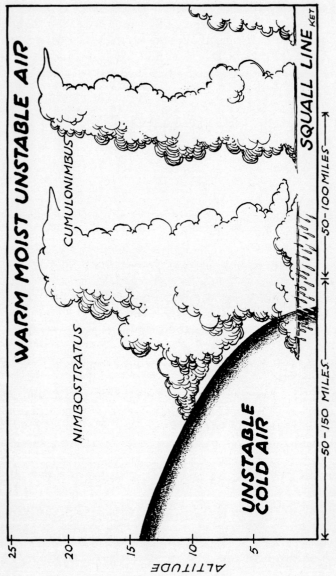

Cold Front: The leading edge of the cold front is steep and preceded by squall lines of cumulus and cumulonimbus clouds—rain and thunder.

7. Fronts: Cold, Warm and Occluded

It has already been mentioned that when a cold air mass and a warm air mass meet, they don't mingle. Therefore, in the mid-latitudes, there is a constant struggle between the cold and warm air masses. Cold air, being heavier than warm, slides under the warm air; or, the warm air can overtake the cold air from the top. As the cold air moves on, it can be replaced by the warm air.

The line of demarcation between the two masses is a sloping one, in profile, and is called a *front*.

A hazard to flight is the sloping frontal surface that brings turbulence, thunderstorms, icing and precipitation. The "leading edge" of a front tells the extent of the weather involved. A steep slope of the frontal surface produces a narrow band of clouds and heavy showers. A shallow slope causes a wide band of fog and light continuous rain.

There are three general types of fronts—cold, warm and occluded. Each is distinctive, so that you will know what kind of weather to expect. Briefly described, a cold front is one in which cold air moves under to replace warm air at the surface.

A warm front is one which has little or no motion.

An occluded front is one in which a cold front overtakes a warm front.

Any one of these can be a "stationary" front, which means it has little or no movement.

Fronts are ever-present in one form or another and have more effect on your flying than anything else nature offers.

Chilly Moments on a Cold Front

More than one weather expert suggests that in the face of a cold front, a pilot is better off on the ground. After this personal experience, I was no longer disposed to argue.

I dropped off my companion at Ottawa, Canada, and took off for Toronto, a flight of some 200 miles. We had already noted the rarity of airports in this country. Our flight had taken us from Burlington, Vermont, to Montreal. From Montreal we had flown up-river to the fascinating, quaint city of Quebec, then retraced our course to Montreal and on to Ottawa.

As I moved along my route over flat green country I began to look ahead with squinting eyes. Was it or was it not a long gray wall highlighted intermittently with bright spots? Soon, it was all too obvious that I was headed for a solid line of weather, a front that was preceded by individual rain showers.

An unfriendly rain storm like that encountered en route to Toronto. Evasive action is to go around the storm by heading for the light spots . . . in this case, to the right. (Photo by Don Downie)

My finger moved down the course on my chart, If I turned left, I would inevitably hit the shore of Lake Ontario. This, of course, would be an easy aid to navigation that would lead me to Toronto. But it was the long way around.

I studied the map further. The nearest airport was quite some distance ahead and slightly to the left of course . . . Oshawa. By this time I was skirting the worst of the rain, and this made up my mind for me. I followed the "iron beam" (railroad tracks) that headed in the direction I wanted to go.

When it became apparent that this "navigational aid" would lead me into the dark of the storm ahead, I looked at the map again. A road led to the left at a right angle to my "iron beam" toward the lakeshore. Just short of the lake another road to the right, and another right angle, led into the city of Oshawa. The airport was on the northwest side of town.

As soon as I spotted the road that led to the lakeshore, I made a left turn and followed it. I kept looking at muddy fields beneath me, hoping against hope that I woudn't have to make a forced landing in one of them. It would be a difficult task to get the plane out again. Besides, it would be a muddy walk and the shiny plane would get filthy. As these thoughts passed through my mind, the next road intersection appeared. In the wet and darkening sky I turned right and moved down, a little closer to the ground. I could almost see the whites of the automobile drivers' eyes.

At last, the murky outline of the small city's skyline appeared. I headed for it like a homing pigeon. Steering just to the right of the city center, I strained my eyes for sight of the airport. Ah, there it was! I made a bee-line for it and landed without benefit of formal traffic pattern. There wasn't a soul in sight. I taxied to a large hangar that was closed. An attendant ventured outside, and I asked if the plane could be hangared.

After the big doors were closed with the plane safely inside, the attendant enlightened me that I had sneaked in under the leading edge of a cold front that extended for more than 100 miles from one end to the other. An RCAF pilot had explored the length of the front in an attempt to get around it and had landed at Oshawa, also to wait it out.

This was a fast-moving front with a near vertical leading edge. If I had departed Ottawa a half-hour later I would not have made it into Oshawa. I might very well have had a muddy walk through some farmer's field—if I was lucky.

Dissecting a Cold Front

The cold front is distinguished by that roughest of clouds, the cumulo-nimbus. The front is usually preceded by a line of such thunderstorms. Other weather accompanying the cold front includes rain, sleet, hail, snow, icing conditions and extreme turbulence. It is not, in short, good flying weather.

The cold front is the leading edge of a huge polar air mass (cold), known as a high pressure cell. It revolves slowly in a clockwise direction over the northern hemisphere.

As the air mass moves away from the polar region and comes in contact with the warmer air of the mid-latitudes, the cold air mass, which is heavier than the warm air, pushes under and causes the warm air to rise. As the warm air is forced upward by the cold front moving in under it, moisture in the warm air condenses and clouds are created.

The cold air mass can move slowly, making the frontal slope shallow and wide and causing less severe weather. If the cold air mass moves rapidly, then the frontal slope is steep and narrow, resulting in severe weather.

The worst kind of weather is found along the cold front in thunderstorms that are referred to as "line squalls."

Weather accompanying a cold front doesn't last long because the front is not very wide or deep. It can be hundreds of miles long, but is scarcely more than 50 miles through. So, it's better

Heads up as one plane scurries to shelter and the lineman stops his work to look at the onrushing thunderstorm dispensing lightning as it rapidly moves over the Springfield, Missouri, Airport. This was part of a squall line preceding a cold front.

to wait it out on the ground. They are rough. The important thing is to recognize one when you see it, or take note when the weather man points out the cold front on his map. If you don't know what one looks like you might think it's going to be friendly.

Recognition of a cold front starts with noticing a south wind and a band of alto-cumulus clouds on the western horizon. Winds may increase and be gusty while the barometer might jump around. And, of course, the presence of thunderstorms or areas of violent turbulence will announce the cold front.

As soon as the cold front passes, the wind will shift to a westerly or northerly direction. The barometer will rise, and the temperature will drop. The overcast that accompanied the front will break into scattered cumulus.

It's now safe to get airborne again.

If you absolutely *have* to fly through a cold front, fly between the storm centers, or over the cloud saddle. If you decide to try flying under, it is recommended that less turbulence and chance of icing is near the ground, or in the bottom third of the ground-to-cloud distance.

Obviously you should avoid the under-the-storm method if you are in mountainous, or even hilly country. Flying between the storms or on top of the saddles is the second best way to make it safely through the cold front. The first choice is always to stay on the ground until the storm has passed.

Warm Front

A warm front is produced when a surge of warm air from the south invades the cold air in the north. It forms a low pressure system.

Winds blow counter-clockwise around the low pressure system in the northern hemisphere. (The reverse is true in the southern hemisphere.) Therefore, winds in the northwest will blow from a southwesterly direction.

The warm southwesterly winds cause warm front weather by blowing up and over the cold easterly winds. When the moist warm winds are lifted by the cold air, clouds form and spread for many miles in advance of the front.

Flying between cumulonimbus from above and . . .
. . . heading around cumulonimbus from below are two routes to safety. (Photos by Don Downie)

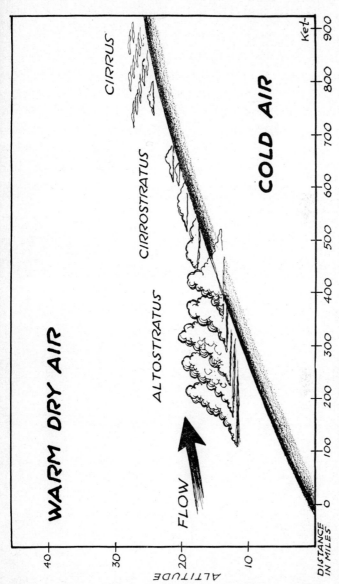

Warm Front: The gradual sloping ceiling shown here, looking from right to left, identifies the warm front. This is the kind of weather encountered in the experience related in the Introduction.

Your first signs of a warm front are parallel bands of cirrus clouds. They are followed by an overcast of cirro-stratus clouds which overlies and leads the solid cloud formations that comprise the front itself. The cirro-stratus cloud deck gives the sky a milky cast and throws a sort of halo around the sun or moon.

Well below the cirrus and cirro-stratus clouds are alto-stratus and alto-cumulus ranging from 8,000 to 15,000 feet in altitude. These are the first cloud banks directly connected with the main warm front surface.

Further into the warm front zone, clouds get lower and lower to the ground. They may even extend all the way to the ground as fog, making landing impossible.

Warm front precipitation can extend as far as 300 miles, and can be preceded by clouds as far as 600 miles or more.

The warm front brings extensive cloud formations and all types of precipitation in the order of rain, snow, freezing rain and sleet.

The pilot should be particularly wary of the poor visibility and low ceiling of the warm front and, in winter, its icing conditions.

If you find yourself flying in snow, your visibility straight down will be reasonably good, but towards the horizon it will be very bad. Should you be in wet snow—unhappy thought—it will cling to your windshield and leave only the side windows clear. Anyway you look at it, flying in snow is even risky for birds.

As you, theoretically, head further into the warm front after encountering snow, you'll run into freezing rain. As this rain strikes your aircraft in cold air, it turns into clear ice. We've already discussed the dangers of icing.

Beyond the freezing rain stage is sleet, which is rain in a frozen or mushy ice form.

The only recommendation here is that upon flying into a warm front and encountering snow, make a fast turn around and head for an airport to wait out the weather in safety.

Occluded Front

Simply, the cold front of dense, heavy air overtakes a warm front of less dense, lighter air. Because the warm front was

OCCLUDED FRONT DEVELOPMENT

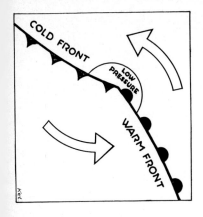

Cold air pressure starts to push into warm air in counter-clockwise movement to Low Pressure center.

Increased cold air pressure starts a bend in the frontal line.

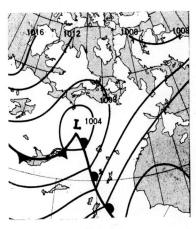

The pressure continues the bend until it reaches the warm front and the front is then occluded. Consider the frontal lines illustrated as an inverted Y, and the tail is the Occluded Front.

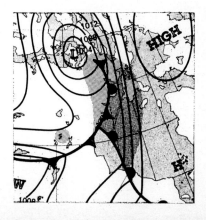

stopped from moving along the earth's surface by the cold front, an occluded front has been achieved.

The occluded front is accomplished in only that one way. And it comes into being because of a cyclone—which is a moving system of winds that rotate counter-clockwise around a center of low pressure.

Cyclone!

What causes a cyclone? Briefly, the cold, heavy polar air produces pressure. The polar front is like a wall that separates the polar air from the warmer air to the south. North of the polar front the winds are easterly, while those in the mid-latitudes are westerly.

When the warm tropical air moves northward at high levels, it piles up behind the polar front's heavy air. Eventually, this piling up of warm air builds up enough pressure so that the polar front is forced to move.

As some of the front bulges southward into the westerly wind stream, the wind is deflected to become southwesterly. This, in turn, causes the neighboring portion of the polar front to retreat north. And this starts a push-pull of cold air and warm air in an apparent counter-clockwise direction. This is the cyclone.

As the cyclone action continues, the cold air pushing through the polar air wall moves into the warm air territory. Being heavier, it pushes under the warmer air and forms a cold front.

Conversely, as the cyclone action continues, the warm tropical air that is pushing northward replaces the cold air that is moving southward and forms a warm front.

In forming the warm front, the warm air follows its native pattern and rises. As it does so, a center of low pressure is formed at the intersection of the cold and warm fronts. This center of low pressure follows the movement of the warm air.

Therefore, the winds of the cyclone (southwest in the northern hemisphere) cause the cyclone to move eastward.

A cold front moves more rapidly than a warm front, and thus it gradually overtakes the warm front. In so doing, the colder and heavier air mass forces the warmer and lighter air mass to move. And it can go only one way—up.

The warm air is thereby cut off from the earth's surface and is called an occlusion.

There are two types of occluded fronts: cold type and warm type. In the cold-type occlusion, the cold front remains on the earth's surface. In the warm-type occlusion, the warm front is at the surface.

The front that is on the surface, whether warm or cold, is the occluded front. In a warm-type occlusion the upperfront precedes the occluded front by 200 miles. In a cold-type, the upper front follows 20 to 30 miles.

The upper front presents a danger to the pilot when it is on his flight path. In the latter stages there is no particular problem.

As the occlusion progresses, the activity connected with it diminishes in severity from its early stages to a gradual dissipation in its later stages. Cloud formations in the upper front change from vertical and flatten into layers during the occlusion process.

No serious flight problems are involved in the later stage at either high or low levels.

Stationary Front

When the movement of the various fronts stops, or when the fronts have dissipated, or the temperatures match on each side of the front, the conditions connected with the front will spread out over a large area and stay put. This is the stationary front, and it can remain this way for several days. With this front you may expect very low ceilings, poor visibility and possibly icing conditions.

This discussion of fronts should certainly convince you that you need the weather station meteorologist. But in the final analysis it is the pilot's responsibility to be wary of these various fronts and to avoid their inherent weather hazards.

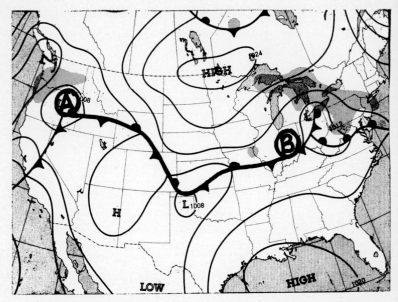

STATIONARY FRONT: Between points A and B the frontal condition has ceased to move. You'll notice that this front is identified by alternating cold and warm symbols on both sides of the frontal line.

8. So You Think You Can Get Out On Instruments !

Famous last words are: "I've flown that route so often I could do it blindfolded." Undaunted by statistics, the pilot gives it a go when the weather itself is the blindfold.

One pilot tried his luck more than once, to help defeat statistics. He lived in the San Francisco Bay area where the typical morning "fog" (low stratus) of summer is usually shallow. It *is* annoying to have to wait hours, sometimes, for the thin gray mass to dissipate in order to take off.

Of course, instrument-rated pilots do not have this problem, but our pilot was not instrument-rated. However, he had developed a rather unique "instrument" of his own. Knowing that the "fog" conditions were usually not deep, our pilot would take off and at full throttle dive slightly under what ceiling there was, then pull back on the wheel and sort of zoom up through the overcast. He made no secret of his successes. Nor was it any secret the day he tried his routine only to find that he couldn't out-zoom the cloud layer. That was the day he became a statistic because of his super-ego and lack of training.

Sky Cover

The less daredevil but more practical pilot will determine from the amount of cloud coverage whether or not he can take off. The term cloud cover, or sky cover, is used by the meteorologists at weather stations to mean a more or less specific description of just how much of the sky is or is not obliterated. Their terminology follows:

    Clear—Total sky cover less than one-tenth

    Scattered—Sky cover from one-tenth through five-tenths

Stratus, a condition similar to that through which the "zoom" pilot lucked out. (Photo by Don Downie)

 Broken—Sky cover from six-tenths through nine-tenths
 Overcast—Sky cover is more than nine-tenths
 The next step is visibility and how it is measured.
 Visibility is sometimes a nebulous thing. For instance, an executive aircraft had completed a 300-mile VFR flight only to be prevented from landing because of less-than-minimum visibility at the airport destination. The pilot was surprised by this information because, as he told the tower operator, "I can see the runway." The tower operator was not sympathetic because from where he sat visibility was less than VFR landing minimums. A half-hour of circling the area and the executive plane pilot had waited out the weather and was able to land, miffed though he was at the delay.
 The pilot in this case was telling the truth. So was the tower operator. Oftentimes ground can be covered by a layer of fog, or haze, which is only a few hundred feet deep. It is quite possible to see the airport clearly from the aircraft—looking down—but the observer on the ground—looking horizontally—may only be able to see a few hundred feet, or less.

Lifting and dissipating fog over San Francisco International Airport is a parallel situation, in a later stage, to that encountered by the pilot who could see the runway and the tower operator who could not.

A typical condition in summer is low flying fog over hills on the east side of San Francisco Bay. (Photo by Keith Dennison)

If the pilot were permitted to attempt a landing in this situation, he might well discover his visibility would be zero when he got close to the ground. He certainly would not be able to accomplish a safe landing under the circumstances, and many accidents have occurred in just this situation.

Fog

Now let's delve a bit deeper into fog.

Fog is a low flying cloud that varies in depth from a few feet to several hundred feet. Its density varies, too, giving visibilities from almost zero to more than 1,000 yards.

From the top, fog can look like a layer of stratus clouds. Unless you have confirmation from a reliable source that it *is* a stratus condition, don't try to get under it. You might find yourself trying to get underground.

Fog is a result of dew point matching the temperature.

To repeat, when the temperature causes the moisture in the atmosphere to condense visibly, the dew point temperature is reached. If we accept the fact that the atmosphere contains water vapor in dry air, then we can understand that the higher the air temperature, the more moisture (water vapor) the atmosphere can hold. The reverse is true, too, if you figure with the same amount of water vapor.

When the temperature goes below the point where it can hold the water vapor invisibly, then condensation will take place and the water vapor will become visible in the form of mist or rain. This same thing happens on a glass containing ice.

Water vapor becomes visible in this damp fog that hides the tops of the towers of the Golden Gate Bridge . . . San Francisco, of course.

You're familiar with the result—a dripping glass. It is caused by the air next to the glass being cooled by the ice to the point of condensation.

There are two ways that the air temperature and the dew point temperature may come together. The dew point can be raised until it meets the air temperature by addition of water vapor from evaporation of water surfaces, wet ground or falling rain. The second method is achieved by lowering the temperature of the air to that of the dew point by cooling the air through contact with a cold surface beneath. The result is fog.

Basically, there are two types of fogs: *air mass* and *frontal*.

Air Mass Fogs

Air mass fogs are of two general types—advection and radiation.

Advection fog is found when warm air moving over water or land is cooled to the dew point by contact with a cold surface. This fog can be carried a long distance in the airstream.

Radiation fog is formed by the cooling of a land surface on clear nights by heat rising. The air in contact with the cold ground is cooled to its dew point, and the fog which forms remains in the area.

The varieties of advection fog are sea fog, tropical air fog, Arctic sea smoke, steam fog and upslope fog.

Best known and most widespread is sea fog. This forms when warm air moves from land over colder sea, or is wind-blown from warm water to colder water and is cooled to its dew point. This is the type of fog you see hanging along coastlines and hovering over inland areas where the fog has been able to drift during the night until stopped by hills or mountains.

When warm water surface air moves over a cold water surface, a dense fog that lasts for days may occur over a vast area. Such fog is common along the Newfoundland coast and the Aleutian Islands.

Tropical air fog results from the movement of a tropical air mass from a lower to a higher latitude over cooler and cooler water surfaces. This fog, like the Newfoundland coast and Aleutian Island type, is extensive.

Arctic sea smoke and steam fog are products of cold air moving over warm surfaces. The fog formed is most likely to be found in breaks of surface ice in Arctic regions, and over inland lakes and rivers on clear nights, especially in late fall before being frozen.

You didn't know there were so many fogs, did you? Hold on, there are more to come.

Upslope fog forms when the wind carries moist air up a slope and the air cools to the dew point. As the air moves up to lower pressure, it cools as it expands. This is a frequent type fog in the Great Plains region of the United States where there is a slight, smooth rise of ground from east to west, and an east wind blows when the air has high moisture content.

So much for advection fogs.

Radiation fogs are more familiarly known as ground fogs. They usually form in the early morning hours and are densest an hour or two before sunrise. They "burn off" from two to four hours later.

Ground fog is caused by a series of circumstances that start with the air just above the ground being warmer than the air at the surface (actually a temperature inversion). There must be a light wind of from 3 to 5 knots. There must be high humidity the evening before. There must be enough moisture to produce saturation. The terrain must be such that it allows the fog to accumulate. These are a lot of requirements, but they are often met.

Sometimes the temperature inversion occurs well above the ground and results in stratus formation which is more often called "high fog," and is well known in California.

Frontal Fogs

Frontal fogs, the second general classification, are not as frequent as air mass fogs. They are found in front of, with, or behind a front and move with it. The frontal fog will be slight with a cold front, and will remain very briefly in one place. With a warm front, the fog can be quite extensive—offering a hazard to aviation. It is caused by rain falling into the lower cold air where the dew point is raised to meet the temperature. The warm front moves slowly so the fog remains over an extensive area for a long time.

"High Fog" that is really a stratus formation on the southern coast of Chile at Punta Arenas.

How do fogs dissipate? Increased wind is one method of disposing of fog. Heating from below by sunlight filtering through the fog to the land surface is another method (this one is noted for giving sunburns to the unsuspecting swimmer). But there is no guarantee that the fog will blow away or "burn off." If you're in the air, go somewhere else and land where it's clear. If you're on the ground, you might work on your suntan.

Fog gave two new private pilots an unnerving experience when they took off from their home port at Houston, Texas, to attend a breakfast fly-in an hour's flight away. They had been aloft about twenty minutes when they spotted what looked to them like smoke hanging near the ground. The "smoke" covered a rather large area and obscured the ground. The neophyte pilots finally realized that it wasn't smoke but ground fog that they were flying over, and it was as far behind them as ahead—obliterating all reference to the ground.

They were scared, but felt the best way out was simply to maintain their compass course and hope that the fog would give out before their fuel.

They had, of course, estimated their time of arrival at the airport destination. As the ETA drew near the two pilots strained their eyes for a glimpse of the ground. The outcome of their experience may be attributed to "beginners' luck" because just as their watches indicated that the ETA had arrived,

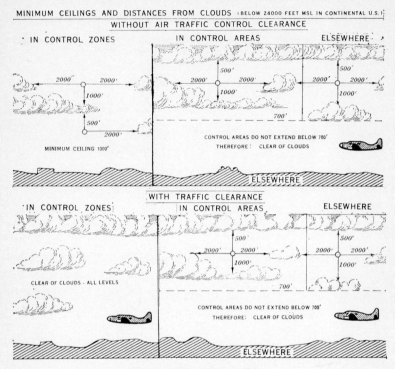

Visual Flight Rules (VFR)

a hole appeared in the fog below and revealed their destination. Down they went to a safe finish, but it took considerably longer than the descent for their hands to stop shaking.

Why Collect Matchbooks?

I was grateful that I was a matchbook collector when the flashlight battery gave out with still 100 miles to go. My companion and I were on our way home after a short vacation in southern California. We were both logging time toward our commercial pilot licenses with still a long way to go before qualifying.

We had departed from Los Banos on the west side of the San Joaquin Valley at dusk and were headed for San Mateo on the San Francisco Peninsula. The little plane didn't boast cabin lights or instrument panel lights, but we had the flashlight ... so we thought.

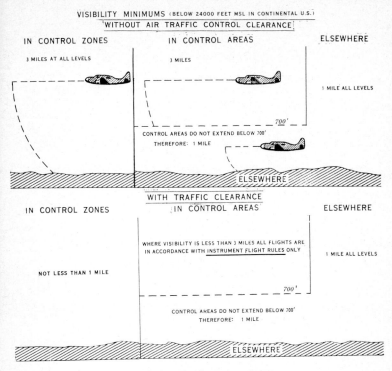

Visual Flight Rules (VFR)

When it became dark enough to need artificial light to read the instruments, we discovered, with a mild burst of hysteria, that the battery-powered torch was of no use. Then I remembered my matchbook collection which saved the day, *and* our nerves.

By the time we arrived at San Mateo, the typical fog condition had set in over the bay, although San Mateo was in the clear. But I still had to fly across the bay and the East bay hills to the airplane's home base.

I took off with the confidence of a 100-hour pilot and made a straight climb out over the bay. I turned in Oakland Radio's low frequency range and listened to the encouraging sound of the call letters. I knew that when I passed over Oakland Radio there would be the "cone of silence" to reassure me that I was on course.

Unexpectedly, at 1,500 feet I skimmed the fringe of the fog layer as I continued climbing. I peered into the darkness for the identifying beacon lights atop the hills behind the Oakland airport. Too soon, it seemed, I spotted a trio of such lights. I scared myself silly thinking that I had flown toward San Francisco instead of Oakland, and that I was looking at lights atop the Twin Peaks in that city. The fog layer below was solid.

Gathering my senses, I told myself that I was on the correct compass course. Oakland Radio was getting louder, I reasoned, which meant I was moving closer to it. But the night was dark and moonless and nothing looked familiar, even when I could see through small breaks in the undercast.

The flight continued with my heart in my mouth. Sudden radio silence helped to restore my confidence. Was I over the cone of silence? By this time I was almost to the East bay hills. The fog had come to a rest against their slopes and extended no further. Finally, I was above the hills and could see lights on the ground ahead and could pick out the familiar highway that led toward my destination. I breathed a sigh of relief. Everything was as it should be. But I was glad when the wheels touched down, and the plane and I were back on *terra firma*.

Visibility and Ceiling

Sometimes visibility is more important than ceiling. Two flight experiences point up this fact. One is the result of knowing the area being flown in, while the other has to do with unfamiliar terrain.

Being a resident of the San Francisco area, most of my flying is done from one of the local fields. The San Francisco Bay is a vacuum for coastal fog. That is, the bay sucks the fog through the Golden Gate at low level until this gray blanket covers most of the water and land around it. The fog is stopped from spreading further when it runs into the low hills that border the bay.

This layer of fog is shallow, and there may be a 1,000-foot ceiling, or less, under it. But in this condition, visibility is often as far as the eye can see.

Local fliers familiar with this condition will often return over the bay to find the blanket cover of fog. But by flying to

Fog seeping through the Golden Gate leaves only the tops of the bridge towers visible as it moves to cover the city of San Francisco at upper left. (Photo by Bill Crouch)

Stopped by the surrounding hills, a layer of fog such as this is discouraging when it sits over your destination. (Photo by Don Downie)

the south end of the bay, they can get under the layer and return to home base. The fog doesn't usually spread beyond the southern edge of the water. Of course, if home base is a controlled airport and ceiling goes below VFR minimums, it won't make any difference how far you can see, unless you declare an emergency.

The other case in point was a flight north through the Salinas Valley, which is near the central coastline of California. Having left the Los Angeles area in scattered rain showers, we flew along the coast leading to Santa Barbara. A ridge of mountains runs parallel to—and practically at—the water's edge. The mountain ridgetop was obscured by clouds.

This same line of mountains begins to slope down as it approaches Santa Barbara. We decided that as soon as there was an opening between the bottom of the clouds and the top of the sloping mountain we'd go across. On the other side, to the north of the mountains, the ground leveled off at a low elevation.

Somewhat beyond Santa Barbara to the west our chance came, and over the hill we went. It was a narrow ridge, and had diminished in height so that it was no longer a mountain. Once across the ridge, the ceiling stayed at about 3,000 feet. Everything below gave evidence of recent rain. It looked fresh and green. We continued north past Santa Maria to San Luis Obispo, where there was another ridge of hills to cross. In this case, the highway wound through the narrow pass, and the clouds descended until they seemed to sit on top of the hills.

On the other side of the inhospitable (in this weather) crossing was the head of the Salinas Valley at Paso Robles. We flew close and took a careful look at the cloud-terrain situation in relation to what the chart showed. It was another narrow ridge of hills with the highway curving through the pass. It seemed safe to proceed. Uneventfully—which was as we wanted it—we made the crossing.

This time the cloud level stayed low. The altimeter indicated around 1,200 feet and we were barely under the ceiling. In places we had to go still lower. But the fact that visibility was probably better than 20 miles made the trip easy and safe.

Haze and Dust

From a distance haze can be confused with fog, which will level off at low levels as well as high. Haze is a combination of "impurities" such as dust, smoke, salt particles and the condensation nuclei that go into making clouds.

Haze is haze wherever it may be . . . in this case near Petropolis, out of Rio de Janeiro, Brazil.

This case of mistaken identity was experienced by a relatively new private pilot who took off from the Lake Tahoe airport, high in California's Sierra Nevada mountains, in sparkling clear air to return to San Francisco. Some 40 miles out, the pilot noted a layer of what looked like fog covering the bay. The immediate conclusion was "it's socked in," and a better-be-safe-than-sorry landing was made at an airport in the clear. The sheepish pilot discovered the case of mistaken identity when another plane arrived from the bay area and clarified the weather situation. It was haze, not fog.

We were four young women on a vacation in Mexico. Two of us were pilots. After a week of sunning and swimming at Puerto Vallarta on the Pacific Coast, we decided to visit the large metropolitan city of Guadalajara, just 130 miles due east.

The interesting note here is that Puerto Vallarta rests at the foot of hills that grow abruptly into mountains to the east. There is a valley that opens onto the small bay, Bahia Bandero, on the south side of which is, or was at the time, the airport. The routine procedure was to take off and circle over the little town until enough altitude was gained to turn on course. We did this, knowing that our new friends would be watching from the beach.

After turning east on course, we spotted far below first one airport then another that were served by DC-3 Dakotas and we wondered how they did it. As we climbed higher and further east, we moved into an extreme haze condition. At 9,000 feet—the mountains there grow high very fast—there was no horizon. We could look straight down and see the ground, but there was nothing to see horizontally except a light brown or sandy colored nothing. It was an uncomfortable sensation, and I coped with my tendency to over-control by letting the plane almost fly itself.

Easing the trim tab forward to put the plane in level flight, we kept our hands off the wheel except for an occasional light touch. By the time we had arrived over the high plateau where Guadalajara rests, the miserable haze was behind us. But the memory lingered on.

Dust Mexican Style

A similar experience, which also involved dust, took place in Mexico on another trip. The desert wasteland that lies south of the California-Arizona border of Mexico never ceases to surprise the flier as he looks down on earth that appears never to have known the footstep of a human—and suddenly sees a pair of automobile tire tracks that seem to lead from nowhere to nowhere.

We were flying over this desolate area on our way to Hermosillo. As we drew close to the city, a yellow-brown wall that was 5,000 feet high confronted us. It was almost like heading into a fog bank, except that, as is usually the case with dense haze, we could still see down. But it was next to impossible to pick up check points. However, there *was* one check point we knew we couldn't miss. It was the highway that ran a more or less east-west course through the city, and extended for miles on either side. When we crossed this road, we would only have to turn left—or right.

The dust was so dense that we couldn't even be sure which side of the city we were on. Everyone cried out an opinion when we crossed the paved road. A right turn was decided on, and after a few minutes without sighting the city, we reversed course and soon saw the airport which is just west of Hermosillo itself.

Fortunately, we did not run into another airplane. But even if there *had been* another aircraft near us, it would have been impossible to spot it. We were decidedly grateful when, the following morning, the dust had disappeared.

Dust that strongly resembles a layer of stratus clouds left good ground visibility in this case in New Mexico. (Photo by Gene Holman)

Towering cumulus clouds seen from the ground at Flagstaff, Arizona.
(Photo by Don Downie)

9. Specialités de la Maison

There are books and more books written on the subject of aviation at the private pilot, or new pilot, level. The authors tell what makes an airplane fly, how to put a plane through maneuvers, and stress the importance of safe operation.

Weather opens another entire realm that the pilot needs to study carefully in order to contribute to his safe operation of the airplane. Too many amateur pilots try to outguess the weather to their own detriment and, frequently, demise.

Up to now, this book has talked of such basics as the various frontal systems with their accompanying weather; clouds, thunderstorms and fog.

There is no shame in knowing your limitations concerning weather, and flying—or not flying—accordingly. A number of aviation organizations are aware that the serious private pilot wants to get as much from his flying as possible, but still doesn't want to get into the training program of the professional pilot.

For instance, the Aircraft Owners and Pilots Association (AOPA) Foundation, Inc., financed an experiment undertaken by the University of Illinois Institute of Aviation. You may have heard of it; it was called the 180-Degree-Turn Experiment. The purpose of this experiment was, per AOPA Foundation instructions, to "Devise simple, practical curriculum for special training program intended only to teach pilot to keep plane upright if caught on instruments, make good enough 180-degree turn to get back to VFR weather or get down through cloud deck."

The University's Institute of Aviation accomplished the mission, and it is now in operation under the auspices of the AOPA. The curriculum is an established training procedure that may be taught by a pilot who has successfully completed the course himself.

In its research for setting up the course, the Institute concerned itself with the causes of serious, near-fatal and fatal accidents which occur because pilots, untrained in instrument flying, attempt to fly under actual instrument conditions.

It was learned that such accidents (1) occur with the greatest frequency in single-engine airplanes of less than 5,000 pounds gross weight; (2) happen in airplanes which are equipped with either partial or full instrument panels; (3) occur with the greatest frequency to non-instrument-rated pilots; (4) happen most often to pilots who have had no previous experience under either actual or simulated instrument conditions; and (5) occur in marginal or sub-marginal weather conditions.

This bears witness to the importance of knowing the meaning of weather indications, knowing your limitations thereto—and acting accordingly.

The publisher of an aviation magazine established the "Live Cowards Club" in August 1961. Sole requirement for membership is admitting to having turned around in the face of weather beyond the pilot's capabilities. The club motto is *"Sodalitas Ignavorum Sempervirens"* which is the club's name in Latin. Membership is open to all pilots of every grade who

In the face of this uninviting view, a 180-degree turn is not to be discouraged.

can certify that they have turned back from a problem which scared them, says founder Robert Blodget.

Of course, the purpose of the club is promotion of flying safety. By making it fun to join the club and profitable to exchange experiences resulting in 180-degree turns, the growing membership is evidence that there is no shame in turning away from a bad situation and talking about it. On the contrary, such an organization only goes to prove that for the intelligent pilot, it is better to be safe than sorry.

Forecasting By Satellite

Sometime in the future there will be weather reports from space satellites that will tell us the weather before it's even made. This will certainly be a boon to the meterologist in preparing forecasts.

It isn't just make believe, either. The Tiros meteorological satellites, and the upcoming Nimbus and Aeros spacecraft, are leading to an eventual world-wide system of long-range weather forecasting. And this will offer far greater accuracy and lead time than at present. Some day it is quite likely that knowledge of the weather will be available weeks or even months in advance.

One of the benefits to come will be knowledge of weather in inaccessible areas where today's weather information is sketchy. The storm-spotting Tiros satellites have already saved many lives and an estimated billion dollars a year in property losses, according to Dr. Francis W. Reichelderfer, chief of the U. S. Weather Bureau.

There is a need to know more clearly how the atmosphere operates. It is hoped that we will learn how the atmosphere behaves as a heat engine by the satellite observations of reflected solar radiation and emitted long-wave infra-red radiation. We know that the sun's heat is the base "fuel" that drives the atmosphere—and creates our weather. We're also aware that the long-wave radiation lost to space is roughly like "exhaust." But we've never been able to see in detail how these losses and inputs vary from time to time, from place to place. With what satellites reveal, the job can start. By applying future similar observations quickly enough, better weather forecasts will be made.

At present, meteorological observations from the ground, and balloons equipped with radios, can provide only a 20 to 30 per cent coverage of weather phenomena, mostly from the Northern Hemisphere and the underside of the atmosphere. Only one-fifth of the globe is covered by any regular observational and weather reporting systems. Extensive areas are not yet covered, and they constitute regions in which storms can be generated and grow without detection before they move over inhabited areas. These gaps may be filled in the next five years by an ingenious satellite system that could photograph the whole panorama of weather, from the march of cold fronts to the birth of storms.

Another future development will be some actual weather control. Cloud seeding is done now to produce rain. The future holds in store the possibility of "engineered" weather of all kinds.

But until these weather cure-alls become realities, the private pilot will have to rely on good sense, training, understanding of what causes weather, weather information provided by professional meteorologists, and, ultimately, his capabilities and limitations. If he does so, chances are he'll live to be the "old pilot" in the verse.

Some Snappy Terminology

If you want to astound your friends with word dropping as applied to meteorology, here are some terms you might remember:

Adiabatic—the word cannot stand alone, but it has to do with the temperature of the atmosphere. You'll find it teamed as:

Adiabatic Process—by which air, when compressed, will have a temperature increase; and air, when expanded, will rise and thus have a lowered temperature. These temperature changes are "self inflicted." That is, there is no outside influence affecting the air.

Dry Adiabatic Rate—assuming the above to be true, there is an established rate of increase or decrease in temperature of $5\frac{1}{2}°F$. per 1,000 feet up or down.

Moist Adiabatic Rate—When air is cooled to its saturation point, condensation takes place. In the process of changing from a

Atop a weather station typical equipment gathers information. (Photo by Don Downie)

gaseous state (air) to the liquid form, heat is given off and is absorbed by the air. This cooling rate, instead of the 5½°F. of the dry adiabatic, is decreased to about 3°F. per 1,000 feet.

Methods of Heat Transfer—Convection, advection, conduction, radiation.

Conduction is the transfer of heat by contact such as air at rest on a colder or warmer surface.

Convection is the transfer of heat by vertical air currents. Air over a hot surface is warmed, and becomes less dense by expansion. Expanded air weighs less and rises. As it does so, there is a compensating downward movement of the cooler air. The up-and-down currents of the warm and cool air are "convection currents."

Advection is like convection, except it is a horizontal transfer.

Radiation is the transfer of heat energy in the form of waves— as in radio or light waves. With only a small amount of matter between the earth and the sun, radiation is the method of heat transfer. The rate at which the earth is heated by radiation (short waves) depends on the absorbing properties of the surface—snow, grass, plowed earth, etc. The rate which the earth cools off at night depends upon the ability of the surface to emit radiation (long wave radiation).

Insolation is the rate at which energy is received at the surface from the sun. It depends on (1) distance from the sun; (2) angle that the sun's rays strike the surface; (3) transmission and absorption of the atmosphere. (About 40% of the sun's radiation is reflected back to space from earth. Dense clouds reflect about 75% of the solar radiation. More heat is received at the equator than at the poles.)

Radiosonde is an instrument for observing the vertical temperature distribution at any particular time and place.

Normal lapse rate is the average rate of temperature decrease or increase in the atmosphere; this is 3½%°F. per 1,000 feet, up or down.

Inversion—where temperature increases with height.

Isothermal or *constant temperature*—where there is no temperature change per 1,000 feet.

Stability—if a parcel of air in the atmosphere is displaced, but in its new position is subjected to forces which tend to restore it to its original position, the parcel is said to be in "stable equilibrium." If that parcel is shunted further from its original position, the atmosphere is in "unstable equilibrium." If the parcel in the new position remains free from forces trying to move it, the atmosphere is in "neutral equilibrium."

CAVU—Ceiling and visibility unlimited.

A Stormy Day for the Weather Girl

On July 13 the daily weather broadcast over the local radio station at Tucson, Arizona, was prepared and presented by Mildred Sprung of the U. S. Weather Bureau.

Just for fun—and so that our pilot readers might have a little better appreciation of weather services rendered—we quote that day's report.

"Inasmuch as our thunderstorm season has begun (we hope!), I think you might be interested in hearing about the duties of a weather observer during a thunderstorm.

"Everything is running smoothly in the office. Nothing to do except to take observations every thirty minutes, plot a weather map, take balloon runs, and answer telephone requests for information ranging from the time of moonrise to the weather from here to Chicago.

"A beautiful cumulo-nimbus cloud, or thunderhead, is forming in the southeast and rapidly increases in size. In a matter of minutes, the entire sky is covered. A distant rumble of thunder is heard—an alert signal for the observer. This, incidentally, requires the sending of a special weather report.

"Soon the clouds darken and lower menacingly over the station. Lightning streaks frequently from one cloud to another as the storm approaches. Soon, a large dust devil comes winding its way toward the station. Windows are closed just a second before the dust hits, knocking visibility to zero.

"Another special report is required for the drastic change in visibility and increase in wind velocity. No sooner is this turned in to the operators for transmission than it begins to rain, and then to hail. Another special report!

"The observer dashes outside to estimate the size of the hail stones—since this is a duty we *must* perform. In the meantime, other employees and airline passengers stand peacefully and comfortably inside—no doubt giving us credit for being as crazy as the weather. The hail stones, in this case, are unusually large, knocking the disheveled observer from one stone to another. The observer reaches the door just in time to take full advantage of a 50-mile gust of wind to shove her in.

"Can't close the door, but that's okay—don't have time for such non-essentials anyway. Another report is sent, taking care, temporarily, of the hail stone situation. The observer 'sighs a heave' of relief, pushes her windblown locks from her perspiring brow, and then telephones the latest reports to the control tower. She then notices a sudden and alarmingly rapid drop in the barograph trace.

"Action! Another special report! A sudden streak of lightning and a deafening crash of thunder fittingly punctuate the previous activities. A quick trip to the roof is made in order to take a thorough check on the weather. Right then is when the cloud opens up and gives . . . just a torrent of water. RAIN, that we've all been wanting so badly.

"The observer makes it downstairs, but not before she is thoroughly drenched. In addition to that, it is now time for a full-fledged observation giving everything from the temperature to the kinds of clouds, their heights and direction of movement. One good thing, it's clear sailing out to the thermometer shelter, as no one else is exposed to the elements. At this observation, the amount of rain is measured. Gee! An inch of rain in a little less than an hour. Phenomenal for Tucson. High speed is maintained since no one relishes the thought of being struck by lightning.

"Everything in weather by this time is coming in large doses. The bare-footed observer, with her war paint streaked and her hair dripping wet, closely resembles a drowned rat. But this is no time for glamour. In spite of the weather and phone calls, the report is completed just in time for transmission.

"And so goes a stormy summer day in the life of a weather observer."